BANTAMS

A Guide to Keeping, Breeding and Showing

BANTAMS

A Guide to Keeping, Breeding and Showing

J.C. Jeremy Hobson

THE CROWOOD PRESS

First published in 2005 by
The Crowood Press Ltd
Ramsbury, Marlborough
Wiltshire SN8 2HR

www.crowood.com

© J.C. Jeremy Hobson 2005

British Library Cataloguing-in-Publication Data
A catalogue record for this book is available from the British
Library.

ISBN 1 86126 786 X

Illustration Acknowledgements
Unless otherwise credited, the illustrations in this book are by the
author. Rupert Stephenson's photographs can also be viewed at his
website, www.rupert-fish.co.uk.

Frontispiece: Blue Orpington female. Photo: Rupert Stephenson

Typeset by Jean Cussons Typesetting, Diss, Norfolk

Printed and bound in Great Britain by CPI Bath

Contents

Introduction and Acknowledgements

Bantams appeal to young and old, male and female, and can be kept virtually anywhere. It is no coincidence that even city dwellers were encouraged to keep a few during World War II – half a dozen can be housed in the space required for two or three chickens, they cause about a third of the damage to a domestic garden or constructed run, and individually eat about half the amount of food. Their character endears them to most people, but they can be particularly rewarding for youngsters, giving them a practical knowledge of biology, livestock management, and even discipline, as they begin to realize that bantam keeping is a seven-day-a-week undertaking. Furthermore, showing bantams is a great leveller, and there is no reason why the little girl with only a pen of half-a-dozen birds at home should not do as well in competition as the millionaire hobbyist with the latest equipment, provided that her pets conform to the particular breed standard.

Situated in a farming area and with a thriving rural studies' group, the library of my secondary school was well stocked with the 'right sort' of books. My personal favourite was a well-thumbed copy of H. Easom Smith's *Bantams for Everyone*, and was a far more interesting read during the maths lessons than the textbooks I should have been studying!

Some thirty-five years later I went into a second-hand and antiquarian bookshop in the hope of finding a first edition of my childhood 'bible', only to be told that as soon as any books of that nature came in, they were almost always already spoken for and were immediately despatched to new owners. It was explained that second-hand poultry

OPPOSITE: Asian male. Photo: Rupert Stephenson

books have always been popular with collectors, many of whom would strip a book for its colour prints. Recently, however, it seems that the old copies have become sought after for the information contained therein, and are firing the enthusiasm of a new generation of poultry keepers keen to discover the pleasure of chickens for themselves. At the top of almost everyone's list is a request for 'anything on bantams', and the owner of the bookshop told me that such is the frequency of these requests that she was becoming a bit of an 'expert' herself, and had, in fact, just installed a small coop and run in her garden for a trio of Silkies.

That's the thing about bantams: they might be the smallest members of the poultry world, but they definitely possess the biggest characters, and once they have you in their clutches, you are a bantam fancier forever. Your personal circumstances might change from time to time, and it may be necessary to spend a period without these busy, bustling birds in your life, but I can guarantee that once the opportunity arises to restock, you will be looking in the 'Animals and Livestock' section of your local paper, and visiting poultry auctions with the desperation of a drug addict looking for his next fix.

You think I might be putting the case too strongly? Well, I and literally hundreds of other bantam keepers can tell you that I am not, and although an addiction to bantams might creep on slowly, the 'habit' soon takes hold. Because bantams can be kept virtually anywhere and with so little effort, it is not always easy to recognize people with a similar obsession; fellow addicts might be living just around the corner without your knowing about them. It might only be when you begin attending meetings of your local club that you realize that you are not alone – even the most unlikely neighbourhood is often home to the occasional sequestered pen of birds.

Bantams are great 'time-wasters': five minutes allocated to feeding and watering can very quickly slip into half an hour as you watch your pen of birds dust and peck their way around the garden while chirruping away to themselves in the manner of ladies doing the rounds of market stalls. To watch them busying themselves with their daily round of activities is wholly therapeutic after a hectic day at work, and just a few minutes in their company helps to put life back into perspective. With a great deal more character than their larger cousins, bantams soon become a part of the family; I knew of one with a regular routine of appearing at the back door each Sunday lunchtime and perching on a kitchen chair to be fed peas from a fork.

Sometimes this obsession with bantams is hereditary, and this was certainly the case in my situation. Brought up on a smallholding and living next door to my maternal grandparents, I spent much of my pre-school life in my grandfather's bantam run, where I was quite happy to sit with my favourite hen on my knee, both of us cheerfully clucking away to each other. The excitement of the discovery of a freshly laid egg was immense, but unfortunately, my eagerness to grab the egg and rush back to the house in order to show off my prize often resulted in a grazed knee and a broken egg, as well as a mass exodus of bantams who would congregate at the entrance knowing that I would probably forget to close the gate to the run.

My own son was no exception: he was given his own pen of exhibition Black Wyandottes for his seventh birthday, but before this he quite often had to be reprimanded for attempting to take one of my own bantams to his bed, obviously considering them to be superior to a stuffed toy – proof, if proof were needed, that bantam keeping is a disease easily transmitted from generation to generation.

The series of Ditchfield 'Little Wonder Books' that I saw during my regular Saturday morning trips to the local pet shop also served to whet my appetite for bantams. Priced at 1s 6d (7p), they were just pocket-money-affordable, and their *Bantams for Pleasure and Profit* taught me much, and inflamed me with a passion so great that a lifetime's obsession was inevitable. If anything I write in this book leaves just one reader with a tenth of that passion, then it will have been a worthwhile exercise.

There are many reasons to keep bantams. It may be the prospect of fresh eggs for breakfast, or just the colour and composition of a particular breed. The exhibition aspect may appeal to the competitive reader, to others it may be the conservation of a classified rare breed.

An *over*-enthusiastic attitude is not, however, a good thing if it encourages the beginner to run before he or she can walk. Heavily feather-legged types such as Pekins or Booted are probably not good first-time choices, especially if it is intended to keep them outside and the run becomes a mud bath. Ideally, these types should be kept on sand or under cover, and as a result could become very time-consuming.

Whatever breed is chosen, their housing will require periodic cleaning if you are not to encourage disease and vermin. The summer months are easy, as the long days give you the chance to clean out pens regularly after school or work – but it is a very different proposition in the winter when weekends may be the only time you see your bantams in anything approaching full daylight.

It is important that any neighbours are consulted before progressing too far with your bantam project. Their attitude may well determine whether you decide to keep a cockerel for breeding purposes, or whether his presence will aggravate the very people you may be relying upon to feed your birds when you are away on holiday. By deciding what you want from your bantams, you are less likely to be disappointed: there is, for example, no point in choosing an ornamental or exhibition variety not known for its laying abilities if all you require is a regular supply of eggs for breakfast!

Although the main purpose of this book is to encourage and advise the newcomer to bantam cocks and hens, the keeping, breeding and showing of bantam ducks is gaining in popularity, and it is for this reason that l have included an extra chapter dealing specifically with their well-being. Much of their requirements are similar to those of bantam 'chickens' but I have, I hope, covered most, if not all, of the obvious differences.

Of course, problems will occur no matter what your eventual choice of bantam fowl, but very few are insurmountable, and perhaps this book will help you avoid a few potential pitfalls until such time as you get to know the local fanciers, some of whom will, I am sure, be only too pleased to assist and advise when asked. That is another great thing about bantams: small in size but big on character, they bring out the best in everyone!

ACKNOWLEDGEMENTS

Being a practical rather than a technical person, first and foremost I must thank Melinda Hall, without whose help and knowledge in computer matters I would have been completely lost. Melinda, thank you for all your encouragement and support, and not just in the writing of this book.

Thanks to Sandy and Eric Compton for their friendship and for 'holding the fort' on the many occasions l have been away from home in connection with research both for this book and for other projects – they are rapidly learning about poultry from the sharp end!

To Francesca Hobson for being my daughter *and* supplying some of the photographs; to her brother Simon, for his help in many ways; and to Steven Thompson for the use of his camera and computer 'trouble-shooting skills'.

No one knows everything, and there is always something to learn when it comes to dealing with livestock. Many have shared their experiences in poultry husbandry in general, and bantam keeping in particular. David Bland of SPR has been wonderfully patient in answering my numerous email enquiries and in checking through some of the chapters. I am immensely grateful to him.

Rosemary Sharpe was kind enough to help me with the 'bantam duck' chapter, both with advice and proof reading. In the same chapter, Colin Murton provided me with further assistance, and also photographs of some of his stock birds.

The following people were enormously helpful with the photography for this book: Martin Clegg allowed me to take photographs of his numerous breeds of bantams, as did the Fuller family; Jack Hughes provided me with slides of his Scots Dumpies; and Phil and Sarah Rant not only let me take pictures of their birds, but kept me well supplied with food and drink whilst doing so. The owners of Fisher's Farm, in West Sussex, allowed me free range on their multi-award-winning farm park, but special thanks must go to Rupert Stephenson for some superb colour pictures (www.rupert-fish.co.uk). I must also thank the David Scrivener Archive for providing photos taken by the late John Tarren.

Liz Fairbrother, editor of *Fancy Fowl* magazine, has been very kind in offering access to material and photographs.

Still others have helped in clarifying various points, amongst them Terry Beebe, David Scrivener, Toddy Hamilton-Gould, Sue Flude and Caroline Hadley. Lorraine Grocott gave invaluable advice on the legalities of planning and environmental health matters. Thanks, too, to Val Porter for allowing me to use some of her research.

Finally, thanks to bantams: without them this book would not have been written!

CHAPTER 1

What are Bantams and Which to Choose?

Towards the end of the 1800s a certain Mrs Ferguson wrote *Hen-wife*, in which she said of bantams:

> These gems of beauty and most treasured and prettiest of pets are, certainly, the most impudent, as well as diminutive, of our domestic poultry. They are ridiculously consequential, and seem as if they pride themselves on their captivating appearance. There are several varieties, all possessing the same passionate temper and, although such perfect pigmies, are most pugnacious, which clearly proves their Javanese origin.

Mrs Ferguson's observations prove that, if anything, bantams were possibly more popular in Victorian times than they are even today. Fanciers of the time appreciated them for their utilitarian properties, and the fact that they required very little space and could, therefore, be kept in small gardens. Bantams were also very much in vogue with the aristocracy, who loved not only their fine plumage and petite appearance, but were also taken with the fact that their eggs contained a much larger yolk-to-white ratio than could be found in the eggs of any large fowl. The Victorians were also immensely fond of exhibiting their livestock, both large and small, and there were many shows organized purely for bantams.

'Gems of beauty ... impudent ... captivating ... most pugnacious.'

OPPOSITE: *Cuckoo Maran female. Photo: Rupert Stephenson*

The heavy breeds of poultry are thought to have no relationship to the lighter jungle fowl.

Although perfectly describing the charm and characteristics of bantams, Mrs Ferguson fails to mention the fact that there is a difference between 'true' bantams and the bantamized versions of large poultry breeds; but before remedying this error, it is necessary to understand a little of the accepted origins of poultry.

A BRIEF HISTORY

It is generally assumed – as did Mrs Ferguson – that all poultry have Javanese origins, and that all stemmed from jungle fowl. Indeed, bantams are thought to be descended from the Bankiva jungle fowl, and were given the name by travellers who discovered them in and around the town of Bantam in Java. At first the name was used to describe just one type, but as the birds began to be exported into Europe, probably some time in the seventeenth century, it became a generic term to describe any kind of small fowl.

While it is generally agreed that the early jungle fowl may well have been the distant ancestor of all game breeds, and certainly had much to do with the light breeds that we know today, the heavy breeds have no such relationship with the jungle fowl, and in fact originated from the Chinese fowls known as Cochin, Brahma and Chabos (Japanese). These birds are known to have been in existence for at least a thousand years. The differences between the two were obvious: whereas the lighter breeds were flighty and roosted high in the trees, the heavier types had large frames, shorter wings, and a complete disinterest in roosting. They preferred to nest in the open, and this, combined with the fact that they were poor flyers, seemed to suggest that they originated from areas with very few predators. It was thought, even as late as Darwinian times, that these heavy birds originated from some scantily vegetated desert, only obtaining water from rivers fed by melting snow. There was – and indeed is – no real explanation, and the only conclusion that even the more investigative naturalists came to, was that what we now know as 'heavy' and 'light' breeds had a completely different ancestry.

Even further back in history are records of domestic poultry being kept by the Egyptians, Greeks and Romans. One school of thought believes that today's breed of Scots Dumpies, ancient enough in its own right, owes its existence to birds brought to

Scots Dumpies are of ancient origin and are thought to have been brought to Scotland by Phoenician traders.

Marco Polo's 'fur-feathered' animals are most surely the ancestors of the modern-day Silkie.

Scotland by Phoenician traders well before the Roman invasion of Britain.

The explorer Marco Polo mentioned 'fur-feathered' animals, which were undoubtedly the ancestors of the modern-day Silkie, and they were also described in a medical book written during the Ming dynasty. Many cures were believed possible by eating both the flesh and the black bones; indeed, some of these beliefs still exist, and thousands of Silkies are bred in both China and Japan each year for 'medical' purposes.

Araucanas, famous for their blue eggs, have a particularly interesting history. Named after the Arauca Indians of Chile, the original breed was rumpless, but birds that, from their description, can only have been Araucanas, were commonly seen around the Mediterranean from about the middle of the sixteenth century. It is probable that traders exporting Chilean nitrates introduced them to Europe, but it is interesting to learn that there are records of hens laying blue eggs in some Scandinavian countries.

Apart from the mention of bantams having come from the town after which they were named, there are few definite records of 'true' bantams until perhaps the mid-eighteenth century, when game bantams were quite common and are often depicted on old paintings, especially those by Dutch and Flemish 'Old Masters'. An exception is the Black Rosecomb, thought by many to be the oldest bantam breed, whose existence was noted by the writers of the day in the latter part of the 1400s.

The 'bantamizing' of large breeds probably began about a hundred years later, but it was not taken all that seriously until the second part of the nineteenth century when the enthusiasts of Mrs Ferguson's era began to set the standards, assisted by the Poultry Club of Great Britain formed in 1877.

Large Light Sussex ...

... and their 'bantamized' cousins.

A Definition of 'True' Bantams and Miniatures

Although Mrs Ferguson neglected to mention it, it is important to define the difference between 'true' bantams and miniatures. Basically, a true bantam is one that has no large fowl counterpart, and is generally considered to be for exhibition or ornamental use, rather than utility. They are in the minority, as most birds available today are dwarfed varieties of large breeds. The latter are sectioned into 'soft feather' (light and heavy), 'hard feather', and 'rare'.

European bantam fanciers tend to speak of 'true' or 'original', and 'miniatures', but sometimes the true bantam is referred to as a 'natural'.

BANTAM BREEDS AND THEIR CHARACTERISTICS

It is not always possible to cover all the varieties of each breed, since in some cases the lists would be immense. Also, some breeds and varieties are recognized in the UK but not in other parts of the world, and vice versa: to be specific about colour, for example, would be open to criticism, depending on where the book is being read. In an effort to collate as much information as possible, I have contacted bantam *aficionados* in America and Holland. In the main, however, this list concentrates on bantam breeds, and *not* varieties of the breeds.

Unless one knows straightaway what breed to start with, it is best to use this list in conjunction with the British Poultry Breed Standards, and to visit several shows before making an eventual choice.

Anconas

The Ancona is one of the miniature bantam breeds, and makes an ideal beginner's bird; it is also a prolific layer of white eggs. Most commonly black, with each feather tipped with white, they originated from the Mediterranean and were introduced into Britain in 1851. The comb is single in the bantam, but rose-comb varieties are sometimes seen in the large fowl. They are classified as 'light' and 'soft-feathered'.

Andalusian Bantams

This is a rare, soft-feathered, Mediterranean breed. It is known as the Andalusian Blue in America: each blue feather is surrounded by black lacing, apart from the sickle feathers of the male and the neck hackles of the female, both of which are black. The comb should be single and of medium size. It is a good layer, but being a light breed it does not make the best of mothers, and it is perhaps best to use a heavier, more reliable type of bantam for incubating their eggs (as would be the case with most of the Mediterranean breeds). In the mid-twentieth century the Andalusian was seldom seen, and British enthusiasts were breeding from just two strains. Thankfully the outlook for the breed is a little healthier today.

The Araucana

Two types of Araucana are recognized in the UK: the Araucana and the Rumpless Araucana. Both types are classified as 'light' and 'soft feathered', and are unusual in that they lay blue/green eggs (the US Standards describes the egg colour as 'turquoise'). The comb is 'pea', and there are no wattles but the face has muffling. There are eleven colour varieties in Britain. Like the Andalusian, the Araucana is a good layer.

The Australorp

This breed gets its name from an abbreviation of 'Australian Black Orpington', and resulted from stock birds imported from Australia in about the 1920s, when it was crossed with the Black Orpington developed by William Cook in the late 1800s. Not only was he famous as the originator of the Orpington breed: by the age of twenty he had written the first edition of *The Practical Poultry Breeder and Feeder*, the poultryman's 'bible', the influence of which is still felt

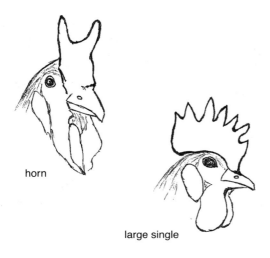

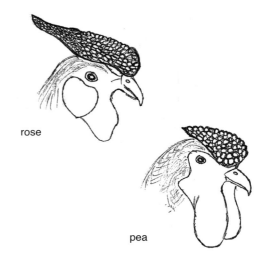

Four examples of comb variations.

today. The Australorp can be said to be both British and Australian in origin, and is classified as soft-feathered and heavy. Eggs are tinted towards brown in colour, and the comb is single and medium in size. Like Henry Ford's motorcar, you can have an Australorp in 'any colour you like, so long as it is black'!

Barbu d'Anvers and Barbu d'Uccle
These are close cousins, and they are both true bantams. Both originate from Belgium, have the same colour variations, and lay an average number of cream-coloured eggs. Both are bearded and muffed, but the Barbu d'Anvers has clean legs and a rose comb, whilst the d'Uccle has a single comb and heavily feathered legs and feet. They require a little more attention in their welfare than some other breeds, especially if one is considering placing them on the show bench. The colour variations and their ornamental stature and size do, however, make them interesting pets, and they have plenty of character.

The Barnevelder
This is another breed imported to Britain from Holland in the early 1920s. A

bantamized version of the large breed, it possesses a single comb and is classified as soft-feathered and heavy. It lays a good number of brown eggs each year, and is 'no nonsense' in its appearance; any of the colour varieties would make a good choice for the first-time bantam keeper. For those who are considering showing, however, it is usually easier to develop one's own strain from a self-coloured variety (of any breed) than it is to do so from, for example, a 'partridge' colouring.

Booted Bantams
These bantams are a rare breed, as nearly all the original bloodlines in the UK were lost at the beginning of the twentieth century. A few dedicated enthusiasts have brought in new stock from Holland and are trying to re-establish the breed to some degree of strength. Like some of the Belgium breeds, they require specialist knowledge in their keeping and, even supposing that stock could be found and purchased, may not be suitable for the newcomer to bantam keeping.

The Brahma
The colours of the Brahma are also varied. Its original name of 'Brahma-Pootra' suggests

16

Both Barbu d'Anvers and their close cousins, Barbu d'Uccle, are 'true' bantams.

that it came from India where there is a river of that name. It reached Britain in the mid-1850s, but was recorded in America roughly a decade before. It is profusely feathered, and is staid and matronly looking in appearance. For a heavy breed it lays a surprising number of light brown eggs each year, approximately 200. The feet and legs are heavily feathered, and the comb is 'triple' or 'pea'. Brahma bantams are miniatures of the large breed, and are soft-feathered.

The Cochin

This bantam is thought to be a miniature of the large breed of the same name. As the Cochin was brought from Pekin in the mid-nineteenth century, it is sometimes wrongly referred to as the 'Pekin'. Some Pekin enthusiasts believe that the first birds were stolen from the private collection of the Emperor of China some time towards the end of the Opium Wars, around 1860. Others maintain that birds imported from China and given to

17

Dorkings.

Queen Victoria sometime around 1830/1840 were bred with further importations, and developed into the breed today known as Cochins.

Whatever the truth, good examples of the breed have hocks completely covered by feathers that, unusually, curl around the joints. The legs and feet should also be covered in feathers. Unlike the Brahma, the comb is small, single and straight, and particularly good specimens will be well serrated. The eggs are light brown in colour and, for a heavy breed, are quite numerous. The Cochin is also classified as being soft-feathered.

There are several plumage varieties, the most common being black, white and lavender.

As with all feather-legged breeds, they require a little more attention than others, but even so they make a good choice for the novice, being attractive and characterful.

The Croad Langshan

This bantam is a miniature of the large breed that was originally imported into Britain as the Langshan by a Major Croad. Originating from China, it is sometimes called the Black Croad Langshan, because although a white strain does exist, it is rarely seen and the black colouring predominates in the breed. It has a single comb and sparsely feathered legs. It lays an annual average of around 160 light brown eggs, and is soft-feathered. The breed standard classifies it as heavy.

The Derbyshire Redcap

The Redcap is unique to Britain; it was derived from the large fowl that was originally produced as a dual-purpose bird, for its eggs and meat. Many fanciers bred birds to emphasize the comb, an accentuated rose comb with a straight leader, developing this shape even further. This breed of bantams is today very rare, but because there is a breed club looking after its interests, it is not classified in the Rare Breeds section, but merely as a soft-feathered light breed.

The Dorking

The Dorking is an old English breed, once again a miniature of the original large fowl. The breed was certainly around at the time of the Roman invasion, and it was mentioned by a Roman writer some time before 50AD. Interestingly, the colour variations of silver-grey, red, white and cuckoo dictate the comb type: examples of the silver-grey and red should bear a single comb, whilst the white and cuckoo must have a rose comb. Another peculiarity of the breed is its five toes and pronounced 'boat'-shaped body. It lays cream-coloured eggs and is soft-feathered and heavy.

Dutch Bantams

The Dutch bantam is the smallest true breed in the world, and the variety of colours available makes them ideal living 'ornaments' in the garden. Like the majority of other true bantams, they are not brilliant layers, but what they lack as egg producers, they more than make up for in character and charm.

The Faverolles

Another soft-feathered and heavy classified breed; the large fowl version originated in northern France where it was bred to be 'dual purpose' – that is, for both eggs and table. It

arrived on British shores in 1886. The bantam miniature has six colour variations, and lays creamy eggs that can sometimes tend towards light brown; it is a good layer. Like the Dorking, the Faverolles has five toes (possibly as a result of the Dorking being crossed into the large breed), lightly feathered legs, a beard and side muffs. Although you will often see the breed referred to as a 'Faverolle', true *aficionados* insist that even a single example of the breed should be known as a 'Faverolles'.

The Frizzle

The Frizzle has always been thought of as an exhibition breed, and although there is a large breed, the bantam variety has always been the most popular. Heavy and soft-feathered, it is the strange feathering that gives the bird its name. In some countries the word 'frizzle' simply denotes a feather type, but in the UK it is a recognized breed, in which each feather should curl towards the head of the bird, and should be as tight and even as possible. Despite being an exhibition breed, the Frizzle lays reasonably well; but those who are thinking of keeping Frizzles should be aware of the fact that the feather formation leaves them unable to cope with particularly wet weather, and so they may well be better housed indoors.

The Hamburg

Although the Hamburg breed came from northern Europe, both spangled and black Hamburgs were bred as large fowl for over 300 years in northern England, where they were known locally as 'Pheasants' and 'Mooneys'. Nowadays, black or white bantam Hamburgs are not recognized as separate breeds due to their similarity to Rosecomb bantams. Colours that are permitted, however, include silver- and gold-spangled, and silver- and gold-pencilled. Classified as a light breed and soft-feathered, the Hamburg is very smart and elegant-looking; it should have a rose comb and white ear lobes. It lays a reasonable number of white eggs each year

Frizzles

(approximately 130), and would make a good breed for the novice. Because they are so attractive they would enhance a garden; however, the spangled variety in particular may become discoloured or 'brassy' if exposed to too much sun. The feathers would moult out, but any suspicion of 'brassiness' would put paid to the current season's show prospects.

The Houdan

In the mid-1950s the Houdan was one of the newest breeds accepted to standard, and caused much excitement amongst the bantam fraternity. One hundred years earlier, the large Houdan had been introduced to England from France, where it was bred as a table bird. It is nowadays classed as a light breed, and is unusual in the fact that it is very heavily crested. Together with the crest, complete muffling surrounds the face, and the comb is 'leaf'-shaped. Like the Dorking and Faverolles, it has five toes and lays white eggs.

Indian Game

Unfortunately, Indian Game are poor layers – but that is perhaps not surprising considering they were originally bred in Cornwall as table birds. The bantam types are heavy and hard-feathered, and have only two colour variations: Indian and Jubilee. Both have pea combs and red ear lobes. They should be very 'chunky' in appearance, and what eggs they do lay are cream/light brown in colour. For someone wanting something a little out of the ordinary, Indian Game are a breed well worth considering.

The Japanese Bantam

This bantam is a true breed with no large counterpart and, unlike some of the other breeds, is unlikely to be mistaken for any other type. Perhaps its most noticeable features are its very distinctive, long upright tail carriage and its short legs. Membership of the Japanese Bantam Club is increasing, which obviously suggests that the breed is becoming more popular. Despite their small size, they lay quite well; the eggs are cream in colour.

The Ko-Shamo Bantam

A miniature of the large fowl, and although it has been known for many years in Japan, its country of origin, until fairly recently it was classified in the UK as 'rare' by the Poultry Club. The fact that it is no longer on that list does not, unfortunately, mean that its numbers are growing, merely that the Ko-Shamo, along with several other rare hard-feather breeds, is nowadays protected by the umbrella of a specialist club.

The Leghorn

Leghorns comprise another breed with its

Maran bantam eggs tend not to be as dark in colour as their large-fowl counterparts.

origins in Italy. Classified as light and soft-feathered, they are miniatures of the large type. Commonly seen in six colour variations, the white Leghorn was imported to the UK via America in the 1870s, and the brown a few years later. They lay a good number of white eggs each year, and are ideal for the first-time bantam enthusiast, despite a tendency to be a little 'flighty'. The cock bird has a large single comb, but to conform to breed standards, the comb of the hen should flop over.

Malay Fowl

Until fairly recently, the Poultry Club officially classed Malay fowl as a 'rare breed'. They are a hard-feather, heavy breed and, as their name suggests, originate from Asia. The large fowl was first seen in Britain in the early 1800s, and for some reason became very popular amongst breeders in Devon and Cornwall. The bantam version is large in comparison to other bantams, and quite leggy to look at. The white and spangled colours seem to be the most popular in the UK, but black, pile and duck-wing are also found. They have a walnut comb and are not particularly good layers; the eggs are cream-coloured. Keeping Malays could be an interesting project for someone who wants something a little out of the ordinary, and you would also be helping to ensure the breed's survival.

The Maran

Sometimes known as the Cuckoo Maran, the large breed is ideal for the newcomer; it is famous for its rich brown eggs. The eggs of the bantam tend to be not quite as brown, but it does lay well, despite being a heavy breed. It is soft feathered. It originates from France, where the breed today often has some feathering on its legs. The comb is single and medium-sized.

The Minorca

The Minorca is very similar in appearance to the Leghorn and, like the Leghorn, is a soft-feathered, light breed. Again, it is a good layer of white eggs. The cock carries a large, erect single comb, whilst that of the hen bird should fold over. Colours are limited to black or white.

Modern Game

This is a miniature hard-feathered breed originating in England, and developed some time between 1850 and 1900 from fighting game fowl. After the abolition of cock fighting in 1949, and with the increasing popularity of showing, judges started looking for taller birds with a shorter hackle and smaller tails. They were originally known as 'Exhibition Modern Game', but the word 'exhibition' was later dropped. The combs when left undubbed would be single and small. Modern Game are similar to the Malay in that they have very long legs. The Modern Game is not a brilliant layer and is a breed that one either loves or hates – I don't think there can be any half measures! Examples of the thirteen colour variations are best seen at some of the larger poultry shows.

The Nankin

The Nankin – not to be confused with the Nankin-Shamo, which is seen in both large and bantam form – is a true bantam, and officially classified as 'rare'. The plumage of the cock is an orange marmalade colour, whilst that of the hen is slightly paler. The comb can be either single or rose, and the eggs are cream in colour.

The New Hampshire Red

Still popular in America where they originated, the New Hampshire Red is also gaining popularity in the UK. It is thought they were bred from the Rhode Island Red without the introduction of any other blood: understandably, this selection process took many years before a standardization could be reached. They are lighter in colour than the Rhode Island, and are classed as heavy and soft-feathered; they would make a good breed for the newcomer, as the single colour is easier to breed to show level. They also lay around 180 eggs during the season, so a pen of them should keep a family in brown eggs for a good proportion of the year.

Old Dutch
This is another true bantam, a pretty bird with five recognized colour variations. Standing very upright and with a single comb, they lay a reasonable number of small cream eggs.

Old English Game
Like the Modern Game, the Old English Game has its origins in Britain, and was tremendously popular as a fighting bird. It is a miniature of the large breeds that were used for fighting, and is classified as light- and hard-feathered. It used to be thought

that 'a good game bird can never be a bad colour', and over the years some thirty colours have been known in the large breed. The accepted variations for bantams today are at the more realistic level of fifteen.

Rumpless Game are sometimes seen, but these are classed as 'rare' by the Poultry Club: available in all the Old English Game colours, their most obvious distinguishing feature is the lack of a 'parson's nose', and no tail.

The Orloff
This bird is doubly unusual, first in looks, and also because it is one of the few poultry breeds originating in Russia. The US Standards recognize three varieties, but in Britain, four are recognized. The Orloff has

Old English with tail …

... and Rumpless.

a pea comb and stands tall on yellow legs. Both sexes have a small muff and beard. They are miniatures, and come under the category of heavy. Also classified as a 'rare breed', examples in the UK would be difficult to find.

The Orpington

The Orpington makes a good beginner's bird. Like a puffball, the four self-colours are equally attractive; they also lay a good number of light brown eggs. William Cook is still known as being the originator of the Orpington breed, and when the black was first introduced at the Dairy Show in October 1886, it produced quite a stir amongst the other poultry fanciers of the time. Within five years of its introduction, the black variety had been joined by the white and the buff, and other breeders were crossing Langshan and Cochin breeds and passing them off as Black Orpingtons, such was the demand. As regards the bantam colours, it was not until the 1950s that whites and blues were developed, and a 'cuckoo' variety has also since come on the scene. The Orpington is soft-feathered and classified as heavy. It has a small, single comb, although the black variety may have a rose comb.

The Pekin

Mentioned earlier alongside the Cochin bantam, the Pekin is, somewhat confusingly, listed in the 'true' bantam section of the Poultry Club, despite being described as a 'miniature of the Cochin' in other works. It was not until the Birmingham show of 1969 that the Pekin breed was accepted in its own right. The breed standards stipulate that it should be single combed, and '… circular in shape when viewed from above, (the) whole outline rounded (with) heavily feathered legs and feet.' There are ten colour variations, and they lay a cream egg.

Plymouth Rock

Although there is a large fowl breed of Plymouth Rock and bantams are therefore miniatures of the type, there is a school of thought that feels that bantams have little or no large breed blood in their evolution. An American breed, much of the colouring in the barred variety is thought to have been the result of breeding with the Scots Grey. In the UK the barred and buffs are perhaps the most commonly seen, but there are four other colours accepted. Classified as a heavy breed with soft feathers, any of the six variations would make a good choice for the newcomer to bantam keeping. They are good layers of cream (US Standards describe them as 'yellowish') eggs.

The Poland or Polish

This breed is a light, soft-feathered miniature. As its name suggests, it originates from Europe, and it is easily recognized by its huge crest, which is neat and compact in the female but more unruly and spiky in the cock bird. It has a small horn comb, and is an excellent layer of white eggs. This is perhaps not the ideal bantam for the inexperienced enthusiast, as it requires a little more attention than the average breeds; but its unusual appearance and particularly attractive colour variations make it worth some consideration.

The Rhode Island Red

This breed can most definitely be recommended to the newcomer to bantam keeping.

They are hardy and happy to adapt to almost any healthy surroundings, and are excellent layers of light brown eggs. They are much darker in colour than their American neighbours, the New Hampshire Red, but are otherwise very similar in stance and formation. The comb can be either single (most commonly seen) or rose comb.

The Rosecomb

The Rosecomb is a true bantam, and is certainly one of our oldest breeds. They are full of character and, being small, would make for ideal pets in a situation where space is limited and where they would have to be confined to a coop and run. (Note that it is always best if the house and run for any breed is movable, or if the occupants can occasionally be given free run of the garden.) They are pretty birds, and probably the colour most commonly seen is black, even though white and blue are becoming increasingly popular amongst the showing enthusiasts. It is believed that they originated from Asia, but their original name, 'Black African', suggests that they arrived in Britain via something of a detour.

The Scots Dumpy

The name of this bantam is sometimes seen spelt as 'Scots Dumpie'; it is another truly old breed, and a miniature of the large breed. In its cuckoo colouring the hen could perhaps be mistaken for a Maran, although the cock bird is much finer and a different shape altogether. Because of their noticeably short legs they have a distinctive waddling gait, quite unlike the walk of any other bantam. The reason for its short legs is traditionally explained by the fact that, in Scotland, the crofts and smallholdings, surrounded by wild countryside in which predators were rife, meant that short legs and a heavy body encouraged it to stay at home. Records suggest that, in many cases, the Scots Dumpy lived in the crofter's house. Although it is accepted as having originated in Scotland, it is thought that their ancestors may have

White Rosecombs.

been brought to Scotland before the Roman invasion.

At one time the Dumpy came in a variety of colours but, due to the loss of genetic material, unfortunately only black and cuckoo are commonly seen today. In fact, had it not been for a dedicated band of breeders in the 1970s, the Scots Dumpy could have died out altogether. The hen is considered to be an excellent broody and mother. They are classified as a light breed, and lay cream-coloured eggs. The comb is single, and the tail of the cock is full and flowing.

The Scots Grey

This bantam was very popular in Scotland up until about the 1930s. Similar to the Scots Dumpy, this light, miniature breed is only recognized as cuckoo in colour, and the breed standard insists on the following characteristics: a 'single comb; white mottled legs, fine, compact, smart bird with well defined markings.' Like the Dumpy, its single comb is accompanied by red ear lobes. Both breeds would make a good choice for the beginner, and the very strong breed clubs can help in locating the whereabouts of your nearest breeder.

The Gold- and Silver-Laced Sebright

Sir John Sebright developed the Gold- and Silver-laced Sebright at the beginning of the nineteenth century, making it one of the oldest British varieties of true bantam. Despite having no counterpart in large breeds, it was thought to have played a part in the production of other laced fowl – notably Wyandottes. Like the Rosecomb, the Sebright makes an ideal choice for the small garden, but can only be expected to lay between sixty to eighty creamy-white eggs each season. Breeding to show standard may be difficult as the correct lacing is most important, but obtaining good quality stock initially will make the development of a successful strain more likely. Enthusiasts have developed a citron colour by crossing the gold and silver varieties together.

Silkies

These are a large fowl, but they are a light breed and have bantam counterparts. They began to be established in the UK sometime during the middle of the nineteenth century, and since then have become extremely popular, in particular when crossed with another bantam type in order to produce the ideal surrogate mother for the hatching of other breeds. Some fanciers feel that there is a danger in using purebred Silkies as broodies because there is a risk of the newly hatched chicks strangling themselves in the very fine feathers. However, having used Silkies myself, I don't think there is any greater risk than there would be in using any other kind of bird. There are five recognized colours, and the bantam can be bearded. The Silkie is often given as a children's pet, and there is no doubt that their unusual feathering fascinates youngsters. They are docile and easy to keep, but because of the feathering, should always have a dry run and warm housing available. They lay a creamy-brown egg and can be expected to lay an average of 105 eggs each year. The comb is known as 'mulberry' or 'cushion', almost circular in shape but preferably broader

rather than long. The male has a slightly spiky crest, whilst the females should be short and neat, resembling a powder-puff. Both male and female have an extra toe at the back of each foot.

The Sussex

Everyone knows the Light Sussex, but the Sussex breed has six other types. Developed from the large breed, the Sussex bantam makes another ideal beginner's bird, but the Light needs to be kept out of sunlight in order to prevent 'brassiness' if it is to be shown. They are excellent layers of cream to light brown eggs, despite being classified as a heavy breed. The plumage variations are, in some types, difficult to achieve to show standard – but that should not put anyone off from trying one of the less common colours. There are always plenty of knowledgeable breeders around who will be only too happy to offer help and advice. The breed has a single comb and a good, solid body, making it full of character and a typical-looking 'farm-yard' bird.

The Welsummer

This Dutch breed takes its name from the village of Welsum, and originally had in its make-up Cochin, Wyandotte, Leghorn, Barnevelder and the Rhode Island Red. It was introduced into Britain in 1929 as a large fowl, and the bantams of today are derived from the large breed imports. For instance, the Welsummer is a good layer of dark brown eggs, and the cock and hen have completely different markings. At a glance, the colouring of the male is similar to that of Black-Red Old English Game, and the markings of the hen could be mistaken for a Brown Leghorn. They have a single comb and are classified as a soft-feather, light breed.

The Wyandotte

This breed is extremely popular, and there are at least fourteen colour variations in the UK alone. Traditionally, the shape of the body should fit into an imaginary circle, and the

The conformation of some breeds such as the Wyandotte and Orpington should be rounded in appearance, and the head, tail and body should fit within imaginary circles.

tail and head into two other circles within the original circle. It is not just the variety of colours or their general appearance that makes these bantams popular, but also their ease of keeping. Like a lot of other pale-coloured breeds, the White Wyandotte needs protection from the sun if you are intending to exhibit your birds, but in spite of this I would suggest that the White is perhaps the best type for the beginner. Many of the other colour variations make for complicated breeding problems, and the Partridge in particular necessitates separate pens of both 'pullet'- and 'cock'-breeding stock. This

'double mating' system is explained in greater length in Chapter Five. The Wyandotte originated in America, and is an excellent layer of cream eggs. It is classified as a soft-feathered heavy breed.

The Yokohama

With its long tail, the Yokohama cannot really ever be mistaken for any other breed. A light, miniaturized version of the large breed, the Yokohama's tail can, in a mature bird, be as long as 60cm (2ft). Despite its exotic looks, enthusiasts of the breed are quite common, and if one is tempted towards something a little different, then it is perhaps an option to be considered. To keep the cock bird in prime condition may, however, be a little difficult, and to maintain a perfect show specimen one would need taller houses and perches than would normally be necessary for other breeds. Laying capabilities can be quite varied, depending on the strain; thus some hens will lay only forty to fifty eggs in a season, whilst others have been seen to average somewhere around one hundred.

It is always possible to pick holes in a list of this nature, but I have tried to cover most of the true and miniaturized bantam breeds around. Some will be commonly found in back gardens and on the show bench, whilst one might go a lifetime and not see a living example of the rarer breeds. America and Europe have their own versions of some of the bantams described, and also breeds that are not seen in the UK, so a truly comprehensive list is virtually impossible to collate, and certainly beyond the remit of this book.

CHAPTER 2

Starting out with Bantams

The beauty of bantams is that, within reason, they can fit into whatever space is available. It is, however, a dangerous assumption to think that just because they are small, they can live happily in a cramped and overcrowded space: they cannot, and even though I have known of bantams being kept in some curious places, without knowledge and some care and attention, problems are bound to occur.

It is the responsibility of any livestock fancier to ensure that their animals and birds are correctly housed, especially in the confines of a built-up area where there are neighbours to consider. The actual construc-tion of the ideal bantam house is covered in the next chapter, but now might be a good time to point out the possible need for plan-ning permission from your local council, espe-cially if you intend building a combined house, veranda, store and possibly even isola-tion/showing pens. Depending on the size of the construction and its proximity to any neighbour's boundary, planning permission may well be required, and it is as well to check out any potential problems before getting started.

A few rats and a noisy cockerel might not pose too great a problem on a smallholding in the countryside, but to allow the same in a

A crowing cock bird is not always a good idea when neighbours are too close.

OPPOSITE: *Spangled Old English Game male. Photo: Rupert Stephenson*

back garden surrounded by urban dwellers is a good way of ensuring a visit from the Environmental Health Authority. It is far better to take preventative steps from the outset than to wait until complaints are registered: once this happens, the bantam owner is legally bound to get rid of the offenders, and if they fail to do so, they could be prosecuted under the Prevention of Damage by Pests Act (1949).

In a built-up environment, whether or not to include a cock bird in your initial purchase of stock will necessitate a great deal of thought. To those interested in poultry, a cock bantam completes the family, and it makes a charming picture as he struts around in charge, offering the best 'tit-bits' to his females, chattering to them all the while. Unfortunately his generally high-pitched crow at dawn on a summer morning when everyone is sleeping with open windows is not likely to endear him to the neighbours, no matter how many eggs are offered as a 'sweetener', and the Environmental Health Authority can insist that measures are taken to prevent such disturbances. To comply, you could try either darkening the roosting area of his housing, or lowering the roof to prevent him stretching his neck to crow. But beware, because if neither of these measures works, an 'Abatement Notice' could be served on you, which would require his removal. Buying only hens is an obvious option, but interestingly, there are on record cases of bantam hens 'crowing'; the possible reasons for this behaviour may well be explained in a letter that appeared some years ago in the *Smallholder* magazine:

> In reply to your crowing hen enquiry: I have to say I have one or two. I have six bantam hens and until this summer, a lovely bantam cockerel called 'Lenny'. Sadly one Saturday afternoon I found Lenny dead in the pen. Since then a couple of hens 'squawked' it out to take Lenny's position as 'top dog', so to speak. They would make an awful strangling crow, practising at any time of the day until you could definitely say it was a cock-a-doodle-do. As winter has approached they have quietened down, but I believe they did it to establish a new pecking order.

It is to be hoped that by studying the breed descriptions provided in the first chapter of this book, and after a few visits to local agricultural shows, newcomers to bantam keeping will have been helped in their choice (the book *British Poultry Standards* will also prove invaluable). It may, of course, be the case that they are happy enough with a pen of assorted birds acquired from friends, in which case the history and provenance of these bantams will be known; but for the rest there is still the difficult problem of sourcing stock birds once the breed has been chosen, and this may take time, even though the pen may be built and the feeding utensils purchased.

SOURCING STOCK

Buying from Breeders

Probably the best way to locate good quality stock is to contact the secretary of the Poultry Club of Great Britain or the relevant breed club. Their addresses can be obtained from magazines such as *Smallholder* or *Fancy Fowl*, and increasingly via search engines on the Internet. Note that when using the postal service, it is only common courtesy to include a stamped addressed envelope with an inquiry, besides which it is more likely to ensure a reply: club secretaries are generally honorary, and neither they nor their organizations can afford unlimited first-class postage.

If your chosen breed is popular in your area – and it is a fact that some parts of the country produce more of one breed than another, sometimes due to tradition, or because local conditions favour one breed over another – you might be fortunate enough to have a choice of suppliers. It is definitely sound advice to visit them all: you will get a better overall picture of the breed, and you

Some of the poultry magazines include comprehensive lists of breeders.

Bantam breeders include all genders and ages!

will find some fanciers to be more helpful than others. Remember that a friend made during the initial purchase will be invaluable in the future should you need further help.

A particular breed may, for one reason or another, be unobtainable within travelling distance. However, as long as you have done enough 'homework' and you have seen enough examples of the breed to know that it is exactly what you want (bearing in mind that pictures in books do not always equate with the bird in real life), it should be possible to contact a breeder by means of the numerous adverts in the poultry press.

It is sometimes possible to purchase good stock at shows and sales. In France, such sales are commonplace.

At the turn of the last century it was common practice to send poultry by train, and while this is less usual nowadays, it is still an option. An alternative would be to inquire if the seller were planning to enter any shows located partway between the two of you, and to arrange to meet there.

Selling Classes

Always use any visit to the poultry section at an agricultural show or to a club event for research purposes and to chat to breeders. It is usually a mistake to buy from a selling class in the hope of picking up first-class bantams, as it is obvious that the best of the stock will have been kept at home. If, on the other hand, all you are looking for is a general representation of the breed and you have no intention of showing, then at least there is the guarantee that the birds will be healthy. There are exceptions of course, and you may be lucky enough to pick up young bantams that are merely surplus to next season's breeding requirements. If you can prevail upon an experienced fancier who happens to be present, he or she may be kind enough to assist you with your choice – but be careful that they are not the vendors, as they could be a little biased!

Poultry Sales and Auctions

Although they are becoming less common, some weekly livestock markets still have a section allocated in which it is possible to buy goats, ferrets, rabbits and poultry. But what you see is what you get, and these venues are not necessarily the best places from which the inexperienced bantam keeper should buy their stock. It is, however, often possible to purchase broody bantams with a mixture of chicks 'at foot' and these could form the nucleus of a pen kept purely for their inclination to sit. There might be little or no come-back on the market organizers should you have cause for complaint, but they still have legal obligations to fulfil. The birds obviously need to be healthy examples of what they claim to be, and so the Trades Description Act

Always carry birds correctly, supporting them with both hands.

is relevant, as might be other regulations' appertaining to that particular area. RSPCA inspectors will be in attendance to ensure that certain offences are not committed. These include:

- Poultry must not be tied by neck, leg or wing.
- Aggressive birds must be caged separately.
- Water must be available in containers that cannot be knocked over.
- If a bird becomes sick or injured during a sale, it must be separated and if necessary, humanely despatched by a skilled, trained person.
- Birds must have sufficient room to stand in their normal position.
- Birds must be removed singly from cages using both hands, and never carried by their legs.
- Birds must be protected from direct sunlight, given adequate ventilation, and be sheltered from adverse weather conditions. They must not be left unattended in their travel containers.

- The containers must be solid sided and allow for adequate ventilation of all birds contained therein.

It is worth noting that these government recommendations apply to all situations (including shows) where poultry is displayed or offered for sale.

Auctions organized by reputable poultry clubs, usually in the spring and/or autumn, can be a good way of buying stock and, like the aforementioned selling classes, will contain healthy, well-produced birds. But beware of letting your heart rule your head, otherwise you may return home with bantams bought on a whim rather than the breed you intended buying. On the down side, birds sold at these venues are usually penned as pairs or trios and will probably contain a cock, so if you were intending to have half a dozen or more birds, this would mean you would be buying unwanted males that would be difficult to re-home. Where space and neighbours allow, you could of course set up two separate pens, each containing a trio, and eventually the cocks could be swapped with the hens to bring in fresh blood to the resultant offspring. However, before going down that route, again it would be sensible to check that both trios have not been put into the sale by the same vendor: if this were the case, the bloodlines would probably be the same.

There are some experienced poultrymen who worry that buying from a sale of this nature encourages the spread of problems such as lice and northern mite; also that it leads to the perpetuation of faults when poor stock is used for breeding by inexperienced purchasers. Nevertheless, provided the organizers are genuine and that sensible precautions are taken (such as are outlined in Chapter 6), I feel they are probably unnecessary worries.

If you already have bantams at home, keep newly acquired birds well away from them for two to three weeks; and if these are the first birds to be purchased, do not buy any more for at least the same period of time.

RECOGNIZING HEALTHY STOCK

When purchasing stock by either of these methods (perhaps with the exception of a general livestock market, where the sellers are unknown), you can be reasonably confident that only well kept and healthy birds are on offer. It is, however, very useful to be able to recognize the general signs of a healthy bantam.

The overall appearance is important, and a bantam should 'look well', with shiny plumage and a full complement of feathers. The eyes must be bright and clear, and the comb and wattles well pigmented with red. They ought to be what is best described as 'waxy' to look at (but not to touch), and typical in shape to the breed standard. It is easier to pick out a good specimen when it is on home ground at the breeders, when it should show itself off and be confident and curious. If the bird is well handled, it will approach and peck at one's feet or even at food held in the hand. Dull-eyed, pale-combed and listless birds that skulk in a corner must be avoided at all costs, and in these circumstances it would perhaps be advisable to make one's excuses and try elsewhere.

Spend some time leaning on the fence and observing the birds' general behaviour: a

Good, healthy stock should have bright eyes, red combs and clean vents.

'happy' bird will be constantly occupied with scratching, preening and dusting. Look out for recently evacuated droppings: these should be dark in colour, firm and well formed, not loose and pale. After making one's final choice, handle the bird and feel its weight – it should be neither under- nor overweight. Inspect under the wings for telltale signs of lice or mites, and check around the vent, which should also be clear of parasites and unsoiled by excrement.

BUYING ADULT OR YOUNG STOCK

It makes sense to buy in the autumn stock that has been hatched in the spring of the same year. By this time, they can be kept as adults and should lay, if somewhat spasmodically, throughout the winter before really getting into their stride the following spring. Spring is always a good time to begin rearing: the cocks are at their most virile, early eggs are most fertile, and the summer months are perfect for the maturing chicks. The best combination is a first-year cock mated with second-year hens (*see* Chapter 5). Young adult birds also have the obvious advantage of being likely to live longer! Commercial poultry fanciers generally keep their stock for only a couple of years before introducing younger and more vigorous replacements; but the whole point of small-scale bantam keeping is to treat birds as pets and part of the family. Just as you would not kick 'granny' out because she is old, you owe it to the matriarch of your flock to reward her egg laying and chick brooding with a long and happy retirement!

It is not uncommon for bantams to live for up to ten years, and I myself had one that survived for fourteen. Although she had long ago given up laying all but the odd egg, she went broody each spring, and always hatched off a sitting of eggs laid by my younger stock.

Occasionally one is offered show bantams that are 'middle-aged', and this is a good and cheap way of getting started, as they will still

Sometimes it is possible to find a breeder with a surplus of youngstock – too much choice can often be confusing!

be capable of producing some youngsters of show potential even though they themselves may be 'past it'. Rosecombs are a good example: mature birds often develop combs that whiten with age and therefore cannot be shown. This is not a hereditary fault, and would not pass down the line to any subsequent offspring.

For those of a patient nature, there is also the option of starting one's bantam empire with a broody hen and the purchase of some fertile eggs for her to hatch. Although this is a slow operation, it has the definite advantage

A broody hen and a clutch of fertile eggs is a good way of beginning with bantams.

of teaching basic bantam husbandry from the very beginning, and is an excellent and hands-on way of learning. However, a major disadvantage is the fact that one cannot predict the sex of the chicks, and you could end up with a pen of cockerels that no one wants. In practice it is unlikely, and anyway cock birds could be swapped for fresh blood, or – hard though it may seem – fattened for the table, or otherwise despatched humanely.

Broody chickens and bantams are still quite easy to obtain. I was fortunate during my early days of bantam keeping in that we lived in a very rural area and had farming contacts who all kept a stock of free-range fowl, some of which always seemed to be broody no matter what the time of year. If ever I found myself with sitting eggs but no broody, it was a simple matter of cycling around the farms until I found one. The usual arrangement was that I borrowed the hen until the chicks were old enough to be taken away, and then returned her to the owner. It is always better to try and find a bantam hen

rather than a large fowl because even though the latter can obviously cover more eggs, they do tend to be clumsy, which could put small bantam eggs at risk.

Hatching eggs can be bought from local breeders, or ordered long distance and despatched by road or rail for a fraction of the cost of adult birds. It might sound highly risky to entrust such a fragile parcel to the Royal Mail or a private delivery service, but several bantam hobbyists to whom I have spoken assure me that it is a relatively safe practice, and provided that the eggs have been carefully packed, there are rarely any breakages.

I would not recommend the first-time bantam keeper to start with an incubator unless they have had previous experience in hatching other forms of poultry or game birds. There are too many imponderables in hatching with an incubator: humidity, temperature, airflow, and a suitable place to keep it are all-important considerations, all of which are done naturally by a broody hen. Of course, there may come a time when an incubator could prove invaluable, and the instructions from the manufacturers, together with the few pointers found in Chapter 5 of this book, will, I hope, help to ensure some successful hatches.

SETTLING IN
THE NEW BIRDS

Carrying cases for the journey home need not be as elaborate as the transport boxes necessary to ensure the safe arrival of exhibition birds at a show (Chapter 6), but they do need to conform with government recommendations as set out earlier in this chapter. They should be rigid and have a solid base, which can be covered with shavings or sawdust to absorb the droppings. Often the best place to position a box so that it cannot slide about is the passenger well at the front of the car, but if you do that, make sure any in-car heating is not blowing directly on to the box. For a short journey there is not too much to worry about, but when travelling for several hours, it will pay to stop and check that the bantams seem happy. Offer them a cup of water – they may not be interested, but if they are feeling

dehydrated they will accept it despite their strange surroundings. Remember to check the box and its contents with all the vehicle doors closed – it would not be the first time that poultry was inadvertently released on to the car park of a busy motorway service station by an anxious traveller checking his stock!

Once arrived home, the new arrivals are best placed into the bantam house with food and water but with the pop-hole closed. After a few hours, or the next morning if you have arrived home in the evening, the pop-hole can be lifted and the inhabitants allowed to explore the run. Do not be tempted to *make* the birds come out: they could become disoriented, and it is far better for them to check out their new home in their own time. Before dusk on the first evening, make a point of seeing that they are on the way to 'bed' and are not having problems in either finding the pop-hole or negotiating the ramp. After dusk, quietly inspect them again and make sure that they are all safely on the perches.

Do not let any of them roost in the nest boxes, because if ever this becomes a habit, it can be very hard to break them of it. Lift

With experience, incubators can give excellent results.

them on to the perch if necessary, and keep doing so until they stay there. Even if you decide to feed indoors (and there is much in favour of this decision, as you will see in Chapter 4), a few handfuls of corn sprinkled away from the house each morning will encourage the birds to investigate all of their surroundings.

In just a short space of time your bantams will be strutting about, scratching and generally giving the impression that they own the place, their charming demeanour reminding you why you chose bantams over any other form of poultry.

Battens or weld-mesh will give newly acquired stock the confidence to negotiate the ramp into the house.

Adding the occasional one or two birds to stock that is already incumbent, so to speak, can be quite difficult. The expression 'pecking order' originates as a result of the fact that, although a cock bird is generally considered to be in charge of his flock, one of the hens will undoubtedly 'rule the roost', and the bringing in of new birds will result in bullying until all are sure of the hierarchy. If possible, introduce all the birds to each other by placing the newcomers in an adjoining coop and run, where they can be seen but are safe from attack behind wire netting. After a while, let them all intermingle – the new 'kids on the block' will probably return to their old roost each evening, but after a while they will join the majority in the main house. Alternatively, when you eventually decide the time is right to put the birds together, do so at night. The next morning, let the bantams out as early as possible. The original inhabitants will head for the open air, and the others will be left to feed and drink in peace.

Contrary to popular belief, it is possible to run a large flock of birds containing two cocks. Ideally they should be introduced together, but a second cock can be included at a later date provided that it is done sensitively and carefully. I would *not* recommend the system advocated by the Victorian breeder William Cook, who in his popular book *Practical Poultry Breeder and Feeder* suggested that cocks could be introduced in the following way:

The heavier one of the two should be put down first, if there is any difference between them. Let him run for a few minutes with the hens, and he will soon commence strutting about as if he is king of the pen. Then the second bird should be put down, and most likely he will be a bit scared at first, but when he sees another cock with the hens, he will perhaps show fight.

After they have had a spar at each other, the smaller bird should be hit on the head, then his head should be put under his wing,

A bantam house of this type is just small enough to be regularly moved about the lawn.

and the bird swung round a few times. When he recovers from this he will feel dizzy, and the other bird immediately knocks him over. In seven cases out of ten they will not fight after this, as the one which is beaten thinks the other bird has completely mastered him.

GETTING RID OF MANURE

It would not normally be necessary to mention the removal of old litter in a book of this nature, as most livestock keepers would have a corner of their land where manure could be placed and conveniently forgotten about, or sold on to enthusiastic gardeners. The whole point of this book, however, is to point out that bantams are such versatile creatures that they can be kept in the smallest of back yards or gardens, provided that potential problems with neighbours are considered alongside the well-being of the birds themselves. In such a situation, the disposal of manure might prove difficult.

A trio of bantams is not likely to produce much in the way of soiled litter and old nest-box material, so it should be a simple matter to offload it on to the local council via the weekly refuse collection. A small pen that is periodically moved over the lawn will have the effect of supplying this natural fertilizer directly to the soil, and the grass will undoubtedly benefit from the nutrients in the faeces. However, care must be taken if young children also have access to the same piece of lawn, in which case it might be necessary to gently rake up the area each time the house and run is moved. For the sake of the lawn, houses should not be kept in one place for longer than a couple of days, otherwise the grass will begin to die off. If there is concern that the lawn may suffer from the bantams scratching the surface, then it should be possible to attach small-mesh wire netting across the base of the run through which the grass will still be available.

Several pens will obviously produce larger quantities of manure, and this may not be so easy to dispose of. Keen gardeners will normally have a compost heap to which small amounts of manure can be added. There is a risk, however, in incorporating too much at

41

one go, as poultry manure – or, more precisely, wood shavings – are slow to break down, and pure droppings, which are rich in sulphate of ammonia, could burn the fibrous roots of some plants if applied too soon or in too large a quantity. Ideally, a compost heap should be constructed in 'sandwich' fashion: a layer of bantam litter is spread over a layer of compost, more compost is added, and then a dressing of soil – and the whole process is then continually repeated. If space allows for manure to be kept for a while, it can be allowed to rot down over the summer and winter months before being dug into the vegetable patch in the spring. Strawberries, onions, potatoes and tomatoes, in particular, benefit from poultry manure, and it can also be quite safely mixed into the flower borders without any danger of burning the plant roots.

It is a shame that peat is no longer considered to be environmentally friendly. At the turn of last century and up to forty years ago when I first became interested in bantam keeping, it was regularly used as housing litter, and although there was always the danger that, left to get dry and dusty, it could encourage respiratory problems, it nevertheless acted as a deodorizer and, being absorbent, helped break down the droppings. It also made the best of garden humus.

Traditionally, gardeners grew some of their finest chrysanthemums, dahlias and many other flowers with poultry manure made liquid by soaking it in a large tub of water and using it to water the flowers.

There are currently on the market, dog 'loos' that are designed to be sunk into the suburban garden, into which 'Fido's' excrement is deposited and chemicals added to break down the contents. If the disposal of poultry manure really is a problem and there are no other alternatives, perhaps it would be worth considering the installation of such a unit.

GETTING HELP

Keeping bantams for pleasure, and eventually breeding from them and possibly showing the youngstock, is really a simple matter of common sense – nevertheless it pays to be armed with as much knowledge as possible. This can come from a variety of

Poultry manure added to the compost heap has untold benefits in the garden.

sources. Health and disease problems are usually best dealt with by a veterinary surgeon – although quite often a fellow fancier will have more experience with bantams than a vet, who generally sees only cats and dogs, and rarely poultry.

To get the most from your chosen hobby, becoming a member of a local poultry club will introduce you to fellow enthusiasts in the area. Do not be put off by the fact that most use the word 'poultry' in their title, rather than specifically mention 'bantams'. Only a handful of such clubs remain in Britain today, but rest assured that the poultry clubs found in any region, and however general their identity, will nearly always have a strong following of bantam enthusiasts in their membership. It will certainly pay to join the Poultry Club of Great Britain, which exists to safeguard the interests of all pure and traditional breeds of poultry, both in the UK and throughout the world. (There is very little point in giving the secretary and address for this or any other particular organization, as they tend to become relatively quickly out of date. The computer will furnish most of these addresses, and has the advantage of always being up to date.)

Likewise, publication titles come and go, but one magazine and one book are definitely worthy of mention here: the magazine is the *Smallholder*, a monthly, which has been established for many years. Within its pages can be found expert advice and column upon column of relevant adverts. The book, which has to be on the bookshelf of every poultry fancier, is the *British Poultry Standards*, published and updated every ten years. The next publication is due in 2007, but the current edition (published in 1997) was the first to include colour photographs, most of which were chosen by the relevant clubs as being good representations of their particular breed. It provides the first step in selecting a breed, and should, perhaps, be purchased and pored over before any final decisions are made.

Just as gardening catalogues provide the vegetable and flower enthusiast with pleasure, knowledge and ideas during the winter evenings, so too will any reading material aimed at the bantam fancier.

CHAPTER 3

Housing

The great thing about bantams is that they can, within reason, suit whatever space you have available. I well remember an example given in *Bantams for Pleasure and Profit* in which the author cited a visit to the seventh floor of a building where three Rhode Island Red bantams were being kept on the balcony. It seemed they were suffering no ill effects and were regular layers – a general indication of good health. This is an extreme case perhaps, but one that proves the adaptability of these little birds.

Bantams do not have to be kept on grass, although they undoubtedly prefer it. If there is no alternative and they have to be sited on hard ground such as paving slabs or concrete, it should be possible to create a run area by laying down pea shingle or even wood bark, both of which are well draining and long lasting. Avoid using fibrous material such as coir, as there is a good chance it may be picked up by the bantams and become stuck in their crop. Likewise straw is not a good idea as it will become wet and matted, and could even encourage disease.

To compensate for the lack of grass, it is doubly important that fresh greens form part of the feeding programme. A definite advantage to the run having a concrete base is the fact that the pen will not have to be moved in

Where grass runs are not practical, wood bark or similar makes a good, absorbent and 'scratchable' alternative.

OPPOSITE: Black Japanese male. Photo: Rupert Stephenson

order to avoid the build-up of disease, and it is a relatively simple matter to sweep up periodically the surface covering and disinfect the hard standing.

A small garden should not be any deterrent to the would-be bantam fancier, and a coop and run of just a few square metres will suffice for two or three birds. It is an advantage to be able to move the run on to fresh grass every few weeks, as it will give the birds interest and help to prevent the build-up of parasites in the soil. Depending upon your personal circumstances, it may be possible to move the run around the garden in the winter, and leave it in one position throughout the summer when the ground is drier and the bantams can spend some parts of the day roaming free.

At the other end of the scale, I know of many situations where bantams are kept by farmers and gamekeepers and are allowed total free range, only returning to roost in the beams of a barn or even the branches of a tree where they are relatively safe from foxes. The one disadvantage to this rural idyll is the fact that the hens tend to lay away from home and it is impossible to find all the eggs. Quite often, the first one knows of a nest is when an obviously broody bantam is seen scratching about the yard, and returns three weeks later followed by a newly hatched brood of chicks. It presents a charming picture, but is of little use to the serious bantam keeper who needs to keep breeding records and to have birds that are easily handled.

For those fortunate enough to have more space and existing outbuildings that can be easily adapted into suitable housing – including storage, show penning and isolation rooms – finance is possibly the only limiting factor. For the majority of people, however, bantam housing is a compromise between the two extremes.

The decision concerning the type of bantam to be kept should be made long before the choice of housing. For a pen of 'mongrels' kept only for their eggs and for companionship, a weatherproof shed and an open run will serve their needs; but if you are intending to exhibit predominantly light-coloured birds, a covered run is essential to prevent the sun causing discoloration in the plumage. The feathers of the neck and back of light-coloured types develop what is known as

Free-range bantams in a rural situation. Charming to see, but not always the most practical arrangement for the new fancier.

Bantam house and run in an ideal position. A south-facing, sunny aspect is favourite. The hedge offers wind protection, but there is a danger that a run devoid of grass could become a mud bath in the winter.

'brassiness', and once this is apparent, there is nothing that can be done apart from waiting for the moult and new feathers to grow. Breeds with extreme features need special consideration: for example, the Yokohama needs plenty of room when perching to accommodate its long tail, which can reach 60cm (2ft) in length, and a great deal of pen management if it is ever to reach show standard; and feather-legged varieties are perhaps best kept indoors on silver sand, shavings or flax and so will require a greater floor area than if they were allowed access to an outside run.

CHOOSING A SITE

Obviously the more space you can devote to your bantams, the happier they will be. Because of the layout of most gardens, it is likely that the chosen spot will be tucked away out of immediate sight of the house. The edges and corners of the garden are probably most suitable – but remember that poultry may attract flies and vermin, which will not endear you to any neighbours. Ideally, there are certain points to avoid and others to include: for instance, bantams can tolerate cold weather, but they dislike wet feet and wind; thus the area needs to be well drained and surrounded by some form of windbreak. If small pens are the preferred option, then the wind is not too great a problem as baffle boards can be attached to the sides facing the prevailing winds. Otherwise, every effort should be made to situate the house and run facing southwards; but if that is not possible, some panelling sections, or even the planting of a fast-growing hedge, will certainly help to keep the bantams warm and happy.

If your land happens to border agricultural property, make sure that any buildings and pens are kept well away from the boundary fence, as farm animals in general and cows in particular seem to take great delight in rubbing against sheds and fencing stakes. For some reason best known to themselves, cows also tend to lick at the most unpalatable of objects, and if they can reach the roofs of any of the poultry houses, they will soon strip them of roofing felt.

Although it may at first seem a good idea to position the houses under trees in order to protect them from the hot summer sun, in practice the buildings will be continually dark and damp, which could encourage parasites to breed. The longevity of the timber will also be shortened. Nevertheless, some form of light shade against the summer sun must be provided, as no type of fowl has sweat glands and they can overheat. On balance, it is better to place the run in the shade and the shed in the open.

OPEN RUNS

It may occasionally be necessary to construct the traditional chicken run, either as an area where bantams can roam freely, or as a perimeter fence inside which smaller coops can be placed. This has the advantage of extra security, which is important if any type of poultry is to be left unattended for any length of time, and there is a risk of predation by foxes or even next door's dog. A pen to keep bantams in will vary slightly in height and construction from one that is built to keep animals out – but either way, it should contain as much vegetation as possible. Grass can always be mown if there is more than the birds can cope with, but a bare mud patch will never be able to re-establish itself unless it is rested for several months. With that in mind, there is much to be said for the creation of a double run, one side of which can be periodically re-drilled with grass seed and rested whilst the other is in use.

Where no grass is growing in confined runs, they should be swept well, especially in hot and dry weather. Even a good grass run will benefit from being swept and rolled two or three times a year, preferably just after heavy rain, as this will make them harder and firmer, making it more difficult for the bantams to scratch up the grass by the roots.

Any grass on which birds are going to run should not be left so long that a white fungus can be seen when you push your hands towards the roots. This encourages a disease called Aspergillosis, which can seriously affect the health of your bantams. (It is more commonly found in damp hay and straw, and is discussed in greater depth in Chapter 7.)

As to the actual construction, the height of wire netting necessary varies with the breed chosen. Some of the lighter Mediterranean breeds are excellent flyers, and may need to be prevented from escaping by the use of an aviary-style run with the top covered by nylon mesh. If they are not intended for show then it is possible to curtail their urge to escape by clipping the primary feathers of one wing. Heavier breeds will not have the same tendencies towards flight, but as a general rule, a fence of between 1.5m (4ft) and 2m (6ft) in height should prevent all but the most determined.

To keep foxes out, it is probably a wise precaution to make the fence of two rolls of netting: a roll with 3.2cm (1¼ in) mesh for the lower half, and one with 5cm (2in) mesh for the upper half. The bottom 23cm (9in) is turned out and pegged down or buried, whilst the top 46cm (18in) is allowed to flop outwards in order to prevent foxes or feral cats climbing up and gaining access. In this situation, the total height of the fencing must be at least 2m (6ft). Extra security is gained by adding an electric fence 23cm (9in) out from the base of the pen, and some 23cm above ground level, with a second strand being placed the same distance above the first. Although initially expensive, a fence of this nature will, with care, last perhaps twenty years, and has the added advantage that there is no need for the shed's pop-hole to be closed every night. When planning the fence line, remember that foxes are quite good climbers, and so avoid including any trees that could offer easy access.

DUSTING SHELTERS

Dust baths are an important part of the daily routine of any bird, and there should be every opportunity for bantams to indulge in this essential activity. Dusting is supposed to help free the bird of any mites or parasites, and it helps to condition the plumage. It is also the

A high fence will prevent even the lightest breeds from escaping and ...

only way in which poultry can cool themselves in hot weather. If a large tree has been incorporated into the pen area they will almost certainly use the base between the roots to create their own dust bath. Similarly, if the house is on a permanent site and is blocked up off the ground by legs or platforms of bricks, as it should be for air circulation and to prevent rats taking up residence, they will use the space underneath in the same way. Ideally, bantams prefer loose sandy earth in which to perform their ablutions, and if such an area cannot be provided naturally, it is a simple matter to create one artificially.

In a permanent run, a dust bath can be made by setting out four posts in a square, the two at the back about 50cm (20in) above the ground, the two forward ones about 1m (3ft 3in) high. Across these are laid spars long enough to support a roof of felted wood or corrugated iron. It is preferable that the shelter faces south, as it then becomes a suntrap and it is also protected from the worst of the wind. If the soil is initially broken up with a fork or spade, the bantams will soon peck and scratch it into a fine powder. The addition of extra dry soil, or better still, a mixture of sawdust, sand and

... if the bottom is dug in, should also prevent access by dogs and foxes.

49

A dusting shelter is easy to construct and will always be appreciated. Photo: Gina Arnold

wood ash taken from the bonfire, is also much appreciated and can be kept in place by nailing scaffolding planks from corner to corner at the base of each post.

For bantams housed internally or in movable runs, all that is required is a wooden box roughly 60cm (2ft) square and about 15cm (6in) deep which is then filled with dusting material. Some experienced poultry keepers recommend the inclusion of a proprietary brand of insect powder sprinkled regularly into the mixture.

HOUSING ESSENTIALS

It seems from looking through the numerous adverts in magazines such as *Smallholder* and *Fancy Fowl*, that there are as many types of house available as there are breeds of bantam. Choosing the right sort can be confusing from such an array on offer, and it will pay to give the special features of each design careful thought before making a final decision. Probably the best way of doing this is to visit any agricultural shows in your vicinity, as many manufacturers will have stands. It should be possible to ask questions, and maybe even to take advantage of a 'special show offer', but the best reason for going is to get some ideas that you can adapt and use in your own construction. Local bantam enthusiasts will be only too pleased to show you their set-up and explain why it works for them. They may even have a unit they no longer use and which they are prepared to sell. I bought my first two 'folds' after calling in on a retired breeder, who sold me the pair for ten shillings (50p). Not having any transport, I dragooned three of my school friends into providing the necessary muscle power, and together we carried them the three miles home!

Some aspects and criteria remain the same no matter what type of accommodation is chosen. Every house should be warm and windproof, and all require light, ventilation, perches and a nest-box. They need to be waterproof, and with a floor surface area large enough for the number of bantams to be kept. The floor should be covered with suitable litter.

Light and Ventilation

Light and ventilation are extremely important. I have heard it said that poultry need twice as much air as humans when the weight-for-weight body-mass ratio is taken into account. Without adequate ventilation,

The first two stages of construction. A house of this size will be suitable for between six and ten bantams.

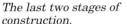

The last two stages of construction.

the air very quickly accumulates carbon dioxide and ammonia and increases the likelihood of respiratory diseases. Windows form the obvious means of ventilating a shed, but actually 'windows' is rather a grand title, as you do not really need the inclusion of glass, just small-mesh wire over which a sliding shutter arrangement is made. Extra holes can be drilled along the top of the front of the house, but these must be fitted with baffle boards in order to prevent a direct draught on

to the bantams as they roost. They are extremely effective in dissipating the build-up of stale air that might otherwise gather in the roof space.

Natural lighting is particularly important if your birds are to be kept indoors for most of the time, because without it, laying and even fertility will be affected. The owner of a French game farm told me that research had shown that the fertility of his cock pheasants improved dramatically when they were given additional artificial light at the beginning of the breeding season, and I have no reason to suspect it should be any different for domestic fowl. Commercial poultry units have long known the value of lighting when it comes to increasing the numbers of eggs laid, and I used to extend the laying period of my White Wyandotte bantams – a breed not best known for their winter output – by the inclusion of a few hours' extra 'daylight' via an ordinary electric bulb. Nature's 'light bulb' is obviously preferable, however, and is also much cheaper than resorting to the National Grid.

Perches

Most birds need to perch. It is a thing they do as a result of millions of years of evolution and for a variety of reasons, not the least of which is protection against predators. Perching also keeps the body warm and off the ground: birds forced to sleep at floor level have been proved to have lower fertility, as the poor conditions affect the ovaries of the female.

Advice as to the ideal measurements for perches varies, but most people agree that a perch should be around 5cm (2in) wide and about 3cm (1in) thick. Softwood is ideal and is best planed smooth, as rough sawn wood is very difficult to keep clean and may cause feet splinters; it is also an ideal hiding place for parasites and red mite. Each bird will require a personal space of around 23cm (9in), and the top of the perches must be rounded off in order that the bantams can get

Perches, lighting and ventilation are very important.

53

a comfortable grip with their claws. The height from the floor depends upon the breeds being kept, but I would be very wary of setting them too high in case injuries result. If I were forced into giving a recommendation I would suggest an average height of about 60cm (2ft), but it really does depend on circumstances. Whatever the height, it is important that all the perches are fixed the same, otherwise there will be a nightly squabble for the highest. It is also better to fix the perches into brackets from which they can be removed and occasionally disinfected, rather than to have them permanently nailed or screwed to the wall.

In some manufactured housing systems, the interior is fitted with perches and a dropping board. In theory, this addition keeps the floor of the house cleaner, but it will necessitate a daily scraping and fresh covering of shavings or sand, and this can be time-consuming, depending upon how many houses one has to clean. A dropping board is, however, a good intermediate point from which birds can access the perches without the risk of damage to themselves.

Perches could also be included in the open run, as the birds like to sit on these during the daytime and preen themselves or just generally watch the world go by. They need only be around 30–60cm (1–2ft) above the ground, but they will be much appreciated, and they will also prevent the birds from crowding into little groups around the run.

Nest Boxes
The best nest boxes I have ever seen were simply wooden packaging crates laid on their

Bantams are not particular where they lay their eggs, but they should be encouraged to use properly constructed nest boxes.

An easily accessible nest box.

side. Originating, I think, from the Egg Marketing Board, they were used to transport trays of eggs and so were almost 'made to measure'. Almost any solid box will do, and I have even seen them made from plastic 25-litre drums, which are ideal, because they can be periodically washed and disinfected.

If they are positioned inside the house, it is a good idea to have them raised off the floor slightly by the use of wooden legs or bricks, but not so high that they become attractive alternatives to the roosting perches. The space below the windows is invariably the darkest point of any poultry house, and is the best place to situate a nesting box as bantams appreciate a dark, quiet corner in which to lay their eggs. The darkness will also reduce the temptation for any individuals to begin

egg eating – not normally a serious problem with busy and otherwise occupied bantams, but one that is difficult to eradicate once started.

In commercially produced poultry sheds, the nesting boxes are often arranged as a bolt-on affair which is accessed from the outside of the house via a single-hinged flap and certainly avoids the necessity of entering the house for the daily egg collection.

One 'nest' for every four hens is about the correct ratio, and it is better to provide too many than too little. If it is decided to build one's own, a box of about 30cm (12in) square is ideal for most breeds. A board of approximately 8cm (3in) running along the base of the front will prevent both eggs and nesting material from being pulled out. Ideally, it

Some old French properties have nest boxes built into the walls of the outbuildings.

should lift out or be hinged in order to facilitate easy cleaning.

The nesting material can be the same as whatever is chosen for the floor, but I would not recommend sawdust for either because of the high dust content, nor would I suggest hay, as it tends to harbour parasites. Interestingly, it seems that bracken harvested and dried in the autumn could be a good substitute, as parasites appear not to like it, for some reason.

Suitable Floor Covering

The importance of suitable floor covering cannot be over-emphasized. Obviously its main purpose is to aid cleanliness and hygiene, and traditionally it would have been straw or wood shavings. In the northern areas of Britain where it was plentiful, bracken was used as the winter bedding of most livestock and also made a good floor covering for the poultry house. Leaves collected in the autumn are another natural alternative and have the advantage of being free, though one disadvantage is they are quite bulky to store until needed. They do, however, make excellent compost when combined with droppings. Chopped wheat straw is also a good medium for general poultry keeping, but is not ideally suited for bantams, especially those kept for exhibition purposes, and other alternatives should perhaps be considered.

Before graduating to our own pens of bantams and poultry, my brother and I used to spend at least a day of every school holidays sweeping up good quality shavings at the local carpenter's shop. Armed with large hessian sacks we would collect enough to ensure our grandfather's poultry sheds were well supplied until the next holidays. Nowadays, a supply of wood shavings is much less time-consuming, and it is a simple matter to purchase compressed bales from the local agricultural merchants. Sold mainly as bedding for horses, there should be very little problem in obtaining white softwood shavings from untreated wood.

Horse owners have a huge choice of potential bedding and there is no reason why any will not do equally as well for bantams. Bales of flax are now available and are extremely absorbent which, after all, is one of the main qualities we are looking for. Slightly more expensive than other material, the extra cost is outweighed by the fact that less is used. Shredded paper is another relatively recent alternative to traditional straw and shavings, and I have used it successfully in both game-bird rearing and as dog bedding. It may have a disadvantage in that newspaper print could discolour white exhibition birds if they were to dust in it too frequently.

Fanciers of feather-legged breeds often advise the use of sand, either on its own or mixed with a little gravel. If kept quite deep it absorbs moisture well and is dry without being dusty. Never be tempted into using sawdust or any peat-based product, not only because of the effect that harvesting peat has on the environment, but also because of the fact that they can be very dusty and, with long-term use, could well affect the bird's respiratory system.

Extra care may be required when it comes to choosing a suitable floor covering for chicks, and what is ideal for adult birds may not suit youngsters. Chicks have been known to strangle themselves in long litter or have even choked by attempting to eat a strand of straw. Wood shavings are probably the safest bet, but they should be of good quality and contain the minimum of small particles, which, if digested, may cause an impacted gizzard.

Whatever the material eventually chosen by the individual, correctly managed, and provided that the worst of the droppings (usually to be found under the perches) are regularly cleared, litter will last a surprisingly long time before requiring a complete change.

TYPES OF HOUSING

No matter what the chosen breed, there is a lot to be said for having a bantam house that is built with a run attached. If at least part of the run area can be covered in order to protect its inhabitants from the worst of inclement weather, so much the better. Such a system allows for regular movement on to fresh grass and is small, compact and easy to manage. For those intending to keep only half a dozen or so bantams, I would say that such a layout is ideal. Combined with a couple of coops and runs in which to house a broody and her chicks, or youngstock, or an adult temporarily 'off colour', it should provide adequate accommodation without turning the garden into a mass of wood and wire.

A house and run with a roof that slopes from front to back is preferred to one with an apex, because unless the sides of the shed are quite high, there is a problem with fitting the perches at a level whereby tail and general feather damage will not result.

One of the best bantam houses I have ever seen was a simple affair built in the manner of an overgrown rabbit hutch. Access was gained via the roof, which was in two halves. Each part slid from centre to end on runners, and allowed cleaning, feeding, egg collection and bird handling to be carried out from above. Corner handles enabled one person to 'walk' the unit on to fresh grass, or two to lift it to a completely new location.

Some combined sheds and runs have the house part off the ground and are similar in appearance to an aviary. As this type of

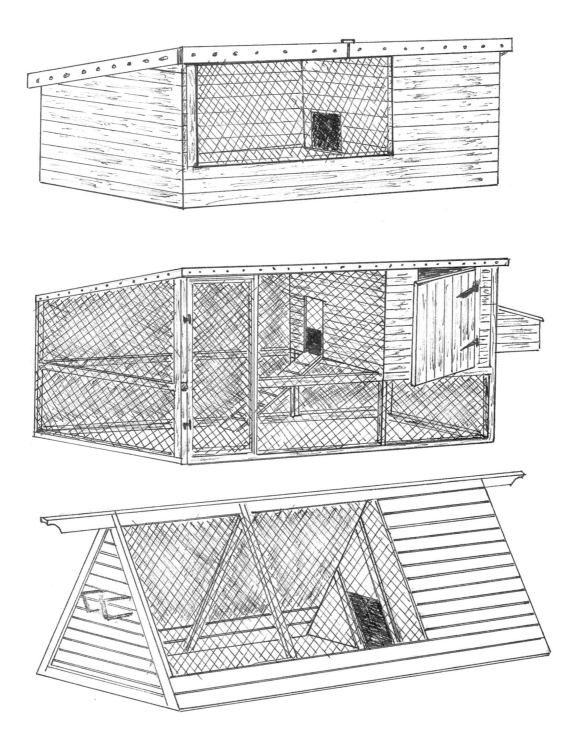

Three types of bantam housing.

accommodation allows the bantams access to the grass underneath, in effect it doubles the available floor area. In such a situation a ramp will have to be provided from pop-hole to ground level. Doors built into the front of this type of housing allow for easier admission to both house and run than would be the case if they were fitted at each end.

If one is lucky enough to have a south-facing wall or a solid fence that forms part of the house or boundary, and where the addition of a bantam house will not offend or hinder family and neighbours, then it may be possible to build a lean-to which combines both house and run. Care would need to be taken into its initial design, as otherwise unforeseen problems could occur. A sloping roof, for example, may run surplus rainwater into next door's garden, even with the inclusion of guttering. Making a waterproof fixing of the roof to the wall or fence is not an easy matter, even for someone with good DIY skills, and it must fit well, otherwise there will forever be trouble with damp in the bantam house. Correctly built, however, it could prove to be an attractive feature, especially if surrounded by screening plants that allow the sun's rays to penetrate, but diffuse any strong winds.

Because the pen cannot be moved, it will be necessary to equip the floor of the run with some well-draining and hygienic artificial base. With a design of this nature it will again prove beneficial to have the roosting quarters about 60cm (2ft) off the ground to protect the wooden floor from damp, to provide additional space for the inhabitants,

Commercially produced show pens.

Even home-made show pens will help in the initial 'training'.

and to discourage rats and other vermin that might otherwise take the opportunity to set up a new home. To avoid the bother of having to enter the run section every time it has to be opened and closed, the pop-hole could be operated from outside the unit by means of a length of nylon and a couple of eyehooks. This simple time-saving device can be included in almost any design and, as I learned from over thirty years of game rearing, will gain the operator hours over a period of months.

Large sheds, either bought cheaply as second hand through the local paper or, if you are extremely lucky, already on your property, can be very easily adapted for use by bantams. Probably the best way of utilizing them is to section the interior into small pens and have a narrow corridor running the length of the shed. Exterior runs can be built, the access to which is via pop-holes cut into the back of the shed. Make them as long as is practicable, and encourage the full use of the run by providing some form of amusement such as scratch feeding, hanging greens or a dust bath at the furthest point. This will help in preventing the area around the pop-hole from becoming a muddy mess.

PENNING SHEDS

Although not strictly housing, the addition of a penning shed will prove useful to the bantam fancier who intends showing his or her birds.

A penning shed is simplest: a building that has good ventilation and light, and is equipped with four or five cages each about 90cm (3ft) square. Depending on the size of the shed, they can be tiered or run lengthways to the room. If you are really serious about showing your birds, it may pay to buy specially constructed show pens in which you can acclimatize the birds you intend to exhibit. Youngstock definitely benefits from being 'trained' in these pens, and will show themselves more confidently as a result. Desirable but not essential to the beginner, home-made boxes will otherwise suffice.

Fitted with a wire front, these pens can then serve a multitude of purposes, rather than just being used for the preparation of show birds. They will, for example, provide a quiet, secluded place in which to install a broody on her eggs, and will subsequently house both her and her chicks until they are old enough to go out into a grassed coop and run. Any new additions bought in, or birds brought back from shows, could be housed here for a short period until it is obvious that they have not contracted disease or are suffering from stress or ill-health. Ideally they should be kept away from your own stock, and their inclusion in a penning shed that already contains vulnerable chicks is not to be advised: but should the shed be unoccupied, it is a perfect halfway house.

A penning shed also provides storage for feed bins, floor litter, medications, louse powder and the carrying baskets necessary to transport birds to and from shows. Useful though it will prove, beware of allowing the penning shed to become little better than a junk room into which everything is thrown, and rats and mice are encouraged.

CHAPTER 4

Feeding

Although a plant will grow virtually anywhere, it needs the right nutrients to bloom and flourish. Carrots will form in clay soil but will be stunted and twisted, whereas if the seed is sown in light, sandy soil, there is a good chance of harvesting vegetables that are straight and perfectly formed.

Much the same applies to bantams. Fed on bread and household scraps they will grow and probably lay reasonably well, but they are unlikely to make exhibition stock or produce healthy and potentially prize-winning offspring. The old conundrum of 'which came first, the chicken or the egg?' applies very well in this situation, in that a malnourished hen cannot produce a healthy, fertile egg, and a poor egg cannot produce a chick capable of growing into a prime example of its kind.

It is important that the bantam fancier buys the best possible stock available when first setting out, but then it is equally important that they are fed on the best possible foodstuffs from thereon in. Fortunately, feeding a balanced diet is a simple matter when compared to fifty or sixty years ago. The books of the day were full of advice to both the bantam hobbyist and the commercial poultry keeper. C.E. Fermor, writing in *Good Poultry Keeping*, insisted that:

The Poultry farmer must have a good working knowledge of feeding and nutrition. He should be able to recognise the common ingredients at a glance, and be able to differentiate between a good and a bad sample.

Bantams should be fed on the best balanced feeds available.

OPPOSITE: Silver Dorking male. Photo: Rupert Stephenson

Herbert Howes went much further in his 1946 classic *Modern Poultry Management* stating:

> It is necessary to embody in any mixture correct proportions of protein, carbohydrates, fats and fibre. All these constituents are available in common foodstuffs such as the following: weatings (sic), bran, maize meal, barley meal, Sussex ground oats, extracted soya bean meal, alfalfa meal, fish meal, meat meal, dried skim milk, wheat, oats, barley and maize. Other foods are available but there is no need to go outside this list for general use. It is advisable to become acquainted with the analysis of the various foods in order to make certain that the ration is well balanced.

It must have been almost impossible for the layman to balance his own ration without a university degree in nutritional science, but by the 1960s, revised copies of both books were encouraging new readers to purchase professionally produced mixes, and I have included their salient advice in italics, as the points are just as important today.

> There is much to be said for purchasing proprietary feeding stuffs of **excellent quality from firms of reputation**. (*Good Poultry Keeping*)

> We need not go into the details of making up a balanced diet here as we can buy special mixtures for doing this, but **we must know how to use these mixtures**. (*Modern Poultry Management*)

C.E. Fermor also mentioned the need to be able to 'differentiate between a good and bad sample' and this fact is also still important: badly stored food, either at the manufacturers, retailers, or after purchase by the bantam fancier, can deteriorate and encourage respiratory problems. For the same reasons it is a mistake to use crumbs, mash or pellets after the expiry date, which is

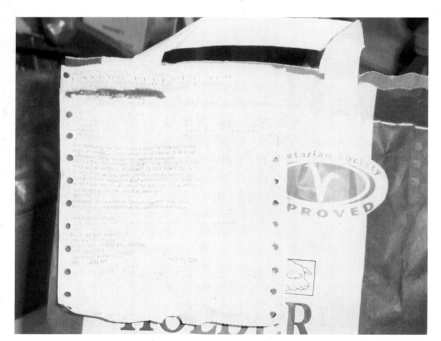

Check the expiry date on feedstuffs – 'out-of-date' products can encourage respiratory problems.

Black Croad Langshans. Photo: John Tarren (David Scrivener Archive)

RIGHT: Ancona male. Photo: John Tarren (David Scrivener Archive)

Lavender Araucana female. Photo: John Tarren (David Scrivener Archive)

ABOVE: *Trio of Black-Red/Partridge Yokohamas.*
Photo: John Tarren (David Scrivener Archive)

LEFT: *Dark Indian Game male. Photo: John*
Tarren (David Scrivener Archive)

BELOW: *Trio of Quail Barbu d'Anvers. Photo: John*
Tarren (David Scrivener Archive)

ABOVE: *Black-Red Malay male. Photo: John Tarren (David Scrivener Archive)*

TOP RIGHT: *Dark Brahma male. Photo: John Tarren (David Scrivener Archive)*

BELOW RIGHT: *New Hampshire Red male. Photo: John Tarren (David Scrivener Archive)*

BELOW: *Buff Sussex females.*

ABOVE: *White-Booted male. Photo: John Tarren (David Scrivener Archive)*

TOP LEFT: *Houdan female. Photo: John Tarren (David Scrivener Archive)*

CENTRE: *Black Orpington male. Photo: John Tarren (David Scrivener Archive)*

BELOW LEFT: *Silver-Laced Wyandotte female. Photo: John Tarren (David Scrivener Archive)*

BELOW: *Gold Partridge Dutch female. Photo: John Tarren (David Scrivener Archive)*

Silver-Grey Dorking male. Photo: John Tarren (David Scrivener Archive)

Black-Red Welsummer male. (The cock has black/brown mixed breast feathers, not solid black as in other 'black-red' types.) Photo: John Tarren (David Scrivener Archive)

Trio of Salmon Faverolles. Photo: John Tarren (David Scrivener Archive)

Mallard Decoy ducks. (The one on the left is an Apricot.)

Silver-pencilled Wyandotte female.

Pair of Cuckoo Marans. Photo: John Tarren (David Scrivener Archive)

(Left to right) Lavender, Cuckoo and Columbian Pekin females. Photo: John Tarren (David Scrivener Archive)

ABOVE: Silver-Spangled Hamburgh male. Photo: John Tarren (David Scrivener Archive)

RIGHT: Barred Wyandotte male.

Silver Sebrights.

ABOVE: Gold Sebrights.

LEFT: Trio of Welsummers showing female feathering.

BELOW LEFT: A Silkie cross-bred hen often makes a good 'broody'.

BELOW: Cross-bred male, but just as handsome as any pure breed.

always shown on the ticket attached to the base of the bag. Where possible, store all floor litter, nest-box material and especially foodstuffs in a cool, dry and well-ventilated space protected from both dust and vermin.

BASIC RATIONS FROM DAY-OLD TO SIXTEEN WEEKS

The sooner you can get any newly hatched chicks interested in food, the better chance they have of surviving. Chicks hatched by a broody hen cause very few problems, as the hen will pick up the crumbs provided and drop them in front of her charges, encouraging them to eat with a few clucking noises

Because they have always been readily available, I have generally fed my young bantams on game-bird crumbs, which have a higher protein level than ordinary poultry crumbs, although the latter will provide the chicks with all the necessary requirements. If deciding to use a game feed, it should be possible to wean the young birds from crumbs to growers' pellets at about three weeks, because the chicks are able to cope with the smaller pellet. It is important to choose both crumbs and growers carefully, as the size varies from one manufacturer to another, and some types of ration produced for commercial poultry can be too large for many bantam breeds, especially the smaller, 'true' bantams such as Rosecombs. With only one brood of chicks to deal with, it is possible to grind up crumbs and pellets if the ideal size is unavailable.

Egg trays make good feeders in the early stages, and game-bird crumbs have a higher protein level than those manufactured for poultry. Photo: Gina Arnold

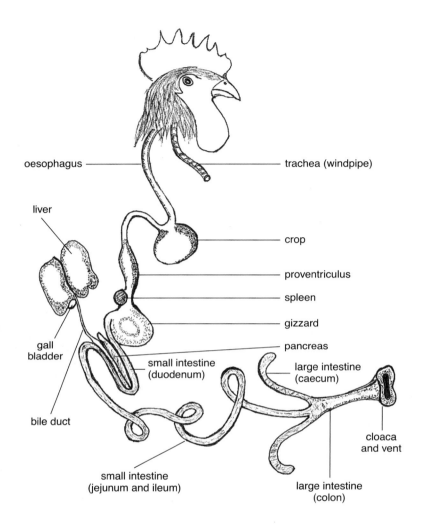

oesophagus ——————

liver

gall
bladder

bile duct

small intestine
(duodenum)

small intestine
(jejunum and ileum)

trachea (windpipe)

crop

proventriculus

spleen

gizzard

pancreas

large intestine
(caecum)

large intestine
(colon)

cloaca
and vent

A basic knowledge of the digestive system is advantageous when it comes to understanding the nutritional requirements of bantams.

The majority of these feeds contain coccidiostats and so can only be obtained from a licensed retailer or direct from the manufacturer; their use is covered in greater detail in Chapter 7. I always include a few finely chopped hard-boiled eggs sprinkled on top of the crumbs for the first couple of days, but I am not sure as to their real value in getting the chicks off to a good start, because I think most are eaten by the broody hen!

As with any young livestock, the aim should be to feed 'little and often', but in practice this is often difficult for most of us to achieve due to our busy daily routine. The best one can do is to ensure that sufficient food is available for the day, though not so much that it gets scratched about by the broody hen and is either wasted or becomes stale.

Although anywhere between four and six weeks is the time given by most manufacturers as the date to change from crumbs to pellets, from my experience it pays to start introducing pellets a little before. I have noticed that chicks seem to be looking for something else in their food after the age of about three weeks, and there is a certain point where, although the food seems to be disappearing, if you look more closely, the crumbs you fondly hope are finding their way into the youngsters' crops are in fact being wasted. Always begin by introducing the growers' pellets very gradually, perhaps one part pellets to three part crumbs.

By the age of sixteen weeks, the bantams should have graduated to layers' pellets or mash, which, although lower in protein, have had other, very necessary ingredients added. If the stock is to be kept specifically for breeding the following year, they can be kept on an ordinary diet until perhaps just after Christmas when breeders' pellets can be slowly introduced. These have additional ingredients that go a long way towards ensuring that the parents are in prime condition and the chicks are born healthy. The occasional clutch of eggs set under a broody just for interest will, however, hatch perfectly well if the parents have been fed on a layers' ration, and it is only when rearing several broods for exhibition or sale that one need consider breeders' pellets. Otherwise continue to feed layers' mash or pellets throughout the year.

The majority of poultry fanciers prefer the simplicity of feeding pellets compared to mash, but others are just as strong in their belief that the birds prefer mash. It is a fact that pellets can be picked up far more quickly than can mash, leaving the birds with little or no further entertainment and the possibility of boredom, which in turn could lead to vent or feather pecking. Nevertheless, even strong supporters of feeding mash admit that it is not a good idea to feed it to show birds because it is messier to eat. As compound feeds, both incorporate everything required by the adult bird; though having said that, the inclusion of supplementary diets of grain and greens will undoubtedly benefit your stock, as well as keeping them active, happy and amused.

Greenstuffs

Bantam chicks kept inside for the first few weeks of their life will do much better if they are given some form of greenstuff. This can be offered at around the end of the first week, but must be finely chopped. Almost anything will do: lettuce, nettles, chickweed, dandelions and young grass will all be pecked at and appreciated. Adult exhibition stock with no grass run will also gain an extra 'bloom' if periodically fed in a similar fashion and, depending on its availability, the collection of suitable material will not add many minutes to the bantam keeper's daily routine. It soon becomes second nature to collect a few handfuls whilst out dog walking, for example; but if that is not possible, one of the best ideas is to sow grass seed in trays at regular intervals. As the grass matures, a tray can be taken into the bantam house where the inhabitants will devour every blade. Once emptied, it can be removed, reseeded and replaced. If a square of small-grid 'weld mesh' is cut to exactly the same size of the tray and placed over the top, it will prevent the soil from being scratched and used as a dust bath.

Whether or not adult birds have a grass run attached to their house, they will still benefit from a diet supplemented by the vegetable garden. The leaves of almost any greens will be scoured for insects before being eaten. If offering whole plants, such as Brussels sprouts, cauliflowers or maize plants, they will keep fresher if they are tied upside down to the fence wire or a hook attached to the side of the shed especially for the purpose. There is nothing more wasteful and unsanitary than to throw greens loose on the floor of the house or run, to be trodden on by the birds. Swedes and turnips can be cut in half and speared on to a nail driven just above bantam-head height, which has the advantage of keeping the vegetable out of the dirt and exercising the bird as it stretches up to peck. Just above head height also ensures that the bantams cannot injure themselves. (The same cannot always be said for the owner though, and I have twice punctured my hand attempting to fix half a swede to a nail. Not to be recommended on a cold, frosty morning!)

A simple framework from which to hang greens. There is nothing more wasteful or unsanitary than throwing greenstuff loose on the floor.

If space allows in the vegetable patch, it might pay to plant a row or two of greens solely for feeding to the bantams during the winter months. In France it is possible to buy 'chou fourrager', a large cabbage-like plant that is sown at intervals during April, May, June or July and can be harvested from September through until February. It is well known amongst French poultry keepers, and is thought to be easily digestible. Why not seek out a packet on your next trip over the Channel and grow it for your birds? I can vouch for its value, as my own poultry go berserk whenever it is on offer. Typically, the French also feed the greens of garlic and onions to their stock, and believe them to have a cleansing effect on the bird's intestinal tract. On the down side, I suspect that these may also taint the flavour of the eggs if fed in sufficient quantities.

Cereal Feeds

The pellets or mash that provides the bird's basic ration will have some cereal in its make-up, but there is no reason why a separate feed of mixed corn should not be given. Usually the pellet or mash is given as a morning feed, and allows sustenance to get quickly into the system. A manufactured compound is digested quickly but because of this, does not 'fill' the bird for as long as wheat or mixed corn. Traditionally, a corn feed is given in the late afternoon in order to ensure the bird is well nourished during the

Bantams will always be pleased to help with any work being carried out in the vegetable patch and, in doing so, supplement their daily diet.

hours of darkness. Providing the grain as a scratch feed out in the run encourages the bantams to range further and work harder at finding their food – though be careful not to overfeed, otherwise the run will resemble a scene from Alfred Hitchcock's film *The Birds*, with wild birds flocking in from surrounding areas to take advantage of a free meal. It also encourages rats and mice, which again is to be avoided.

Young bantams can be fed mixed corn from the age of about ten to twelve weeks, obviously in addition to their pellet ration. At this stage, it is better to provide cracked wheat and kibbled maize, which should be available from your usual supplier. It can be fed whole by the time the bantams reach the age of about sixteen weeks. I would always err on the side of caution when giving maize to

either young or old birds, as it is extremely fattening. I have cut open pheasants that have had free access to maize and found huge fat deposits around their abdomen and ovaries. As with most things, moderation is the key. However, many bantam exhibitors I knew in my youth used maize to great advantage when preparing their show birds. The addition of maize to a bird's diet will improve the leg and feet colouring of yellow-legged breeds and will also intensify the yolk colouring, a useful tip if ever you intend to show a plate of eggs. Some poultry keepers only ever give their birds maize in the winter, purely because its high fat levels help to insulate against the cold. For the same reason, others include a little wild bird seed in the winter diet, although this is only ever practical with a small number

Grain as a 'scratch' feed encourages bantams to exercise, and prevents boredom.

of birds due to the extra cost incurred in its purchase.

A bantam will have a daily food intake of roughly 113g (4oz), and most hobbyists feed a ratio of about 75/25 per cent pellets to grain. There are no hard and fast rules, but it is important that the commercial ration forms the highest proportion of the diet. Twelve to sixteen bantams will eat roughly a total of 20–25kg (50–55lb) of food every three weeks. It is not economic to buy small quantities, but fortunately, pellets and grain are available in sacks of this size from any agricultural supplier. Do not, however, be tempted into buying any larger amounts, as there is the problem of storage and the worry that it may go past its 'sell-by' date.

Cooked Foods and a Wet Mash

Before pelleted foods became easily available, it was common practice to mix a daily portion of cooked household scraps with mash and feed as a moist, crumbly, delicious-smelling complete feed. Additions to the mix might also include biscuit meal and pulses. The birds did well on it; they had to, during and immediately after the war when little else was available. (A daily portion of balancer meal for six hens required six ration books – no wonder bantams were so popular, only

A daily combination of pellets, wheat and maize makes a good balanced diet.

requiring half the amount of food of a large fowl!)

There are still some enthusiasts for the 'cook-and-mix' method, but they are in the minority, mainly because of the fact that it becomes a precise science whereby only as much feed is made up as the birds can eat within about thirty minutes. Food left around for very much longer than this begins to go sour, especially on warm days. There is also the inconvenience of requiring special saucepans for the job, and having to mix at least twice a day. On the plus side, it allows for the best use of household scraps as the necessary cooking makes them more palatable.

FEEDING DURING THE MOULT

Even though bantams generally cease laying during the moult, and the summer show season has finished, they should still be kept on their usual feeding programme. The later in the autumn a moult starts, the longer the whole process will take, and years ago there was a school of thinking that suggested an early moult could be encouraged by feeding only a whole-grain diet at this time. A sudden change in routine will undoubtedly start an early moult but is very detrimental as the effort involved takes up extra energy, putting the birds' health at risk. Provided that good nutrition is maintained, however, there is no reason why they should not emerge from the moult in prime condition. If it is ever necessary to change the diet, the transition must take place slowly over several days to prevent an unscheduled moult and to avoid possible digestive problems.

GRIT

Without the inclusion of insoluble grit in its diet, it is virtually impossible for any bird to make effective use of its food. Grit in the gizzard acts as a grindstone, and makes up for the absence of teeth. The food stops first in the crop before being passed through the glandular stomach and eventually into the gizzard. The presence of grit there aids the natural grinding action of the internal muscles, which assist enzymes and bile secretions as the food travels through the body. Bantams ranging on light, stony ground may well find all they need whilst pecking about, but it is only sensible to provide some extra just to be on the safe side.

Oyster-shell grit is also a very important addition to the diet. It provides an extra source of calcium and is useful in ensuring that the eggshell is as hard as we would expect it to be, rather than soft. Having said that, feeding too much extra oyster-shell in addition to that found in the food mix can upset the calcium/phosphorus ratio and have the opposite effect, causing thin and weak-shelled eggs. The calcium/phosphate balance is unlikely to be upset in normal circumstances, however, and problems only usually occur when bantams are not fed a good ration, just corn, or when they are only given oyster-shell instead of flint grit.

Insoluble chick grit (mainly consisting of granite and flint) is available and should be

Any agricultural suppliers will have an overwhelming supply of supplements and vitamins.

given from about the first week, and then every month of their lives, altering the size as they grow. If this is done, there is little fear of gizzard impaction.

SUPPLEMENTS AND SPICES

I remember being advised to give my original White Wyandottes 'Karswood Poultry Spice', to add potassium permanganate periodically to their water, and cod liver oil to their food. As the advice was given to me by poultry men with many years experience and of whom I was in awe, I slavishly followed it without question.

Even though it is established that manufactured pellets or mash contain all that is necessary for the bantam's well-being, there may be certain occasions when they would benefit from a supplement in either food or water. Multi-vitamins, for example, can be added to the newly hatched chick's drinking water for the first three days of life, and they can also be used whenever a bird is likely to

Bantams should not have to rely on ditchwater – static and stagnant puddles can breed disease.

Make sure that the water vessel is large enough, and never leave it placed in direct sunlight.

experience stress, such as immediately prior to and after a show, or when being transported to a new home. They also have a value when birds seem off colour – although there is always a risk that including it then may mask the real symptoms and delay the detection of a disease. Always remember to change treated water at least once every twenty-four hours.

Pure white fishmeal provides heat and energy, and small quantities can be profitably included in the winter ration in preference to maize. Unfortunately, fishmeal is impossible to buy in small amounts, but one way round the problem could be for a poultry club or smallholders' society to buy a 25kg (55lb) bag and split it and the cost between its members. Both fishmeal and cod liver oil are useful additions to the diet of show and exhibition stock, as they help to improve the 'bloom' on the birds' plumage. Calcified seaweed is often used to promote feather growth either during the moult or a few weeks before the show season begins. The old-fashioned poultry spices with the maker's name included in their title – 'Karswood' and 'Battles', to name but two – were basically minerals, all of which are found in the pellet or mash mix.

WATER

It ought not to be necessary to have to mention water under a separate heading. Common sense tells us that all living things need it, and it is well known that animals can live longer without food than they can without water. Perhaps it is because it is generally taken so much for granted that problems can occur. Water helps in maintaining a steady body temperature, and is a major constituent of blood. It is also necessary in the removal of waste body materials.

Just because wild birds will drink from the muddiest puddle or ditch, it might be thought that poultry do not need a constant supply of fresh water; but providing water that you would drink yourself is a good habit to adopt. If you only see your birds twice daily, make sure that their water container is large enough to ensure that they do not run out on even the hottest day. As a general rule, a 4.5ltr (1gal) drinker will suffice for a dozen bantams. Try and position it in a shaded part of the run: it is surprising how quickly water can heat up when the sun is pitching on to a galvanized container at midday. Water intake can increase by as much as 25 per cent in hot weather.

If feeding mash, keep the water some distance from the food, and watch out for fouling from the mash stuck to the bird's beaks.

Chicks should be given the opportunity to drink clean, fresh, aired water before they first feed. For the first day or two they can live off the absorbed yolk sac, and at this stage water is much more important than food.

If water is readily available, bantams will not be tempted into drinking from other sources which could be contaminated, and thus the chances of illness are minimized.

DRINKERS, FEEDERS AND HOPPERS

Although almost any clean container will do for food and water, some are more suitable than others. A shallow plastic plant pot

A selection of drinkers and feeders suitable for all stages of bantam rearing.

Chick drinkers need to be shallow in order to prevent the possibility of drowning. A small length of hosepipe or pebbles fitted at the base of a deep drinker will help in overcoming the problem.

saucer is ideal for feeding chicks but is almost bound to be turned over by the broody hen as she encourages her youngsters to feed. A heavy bottomed, glazed pot, such as those found in pet shops for rabbits, is more likely to be successful and will incur less food wastage. On the other hand, it is essential that access to the food is made as easy as possible for the first few days. In the days when I reared thousands of pheasant poults, I found egg trays to be the best feeder. They are cheap, hygienic and disposable. The moulding prevents too much food from being scratched out, but at the same time allows them to get right in amongst the crumbs. The tray must be pushed firmly into the wood shavings to avoid any chicks squeezing underneath and getting trapped. After the first week, normal

chick feeders can be used – but remember when using those designed for the poultry industry, that broiler chicks are larger than bantams of the same age, so make sure that the lip or rim is not too high off the ground.

If birds are fed 'ad lib' they will never be sufficiently hungry to all want to feed at the same time, and so it is not necessary to ensure that troughs and hoppers are large enough for all to feed at once. They must, however, be large enough for birds at the lowest scale of the pecking order to have access without fear of being bullied by the others. For this reason, it may pay to have one more feeder than you really need. Two designs that protect food from being wasted are troughs with a central spinning spindle, and tubular hoppers.

Some breeds of bantam – for example, Polish – may well need special feeders and drinkers if damage to the crests of exhibition birds is to be avoided.

If there is no alternative but to feed outside, it is possible to buy feeders protected from the elements by a hood, or alternatively, one could construct a weatherproof shelter and feed under that. The major disadvantage to this system is that the shelter cannot be moved and may become a mud bath in wet weather. Always try and move drinkers and feeders regularly so that the potential build-up of disease in one spot is avoided. Feeders and drinkers in the house will benefit from being raised off the floor by some means so they do not become filled with shavings. With small numbers of birds I always feel that troughs or dishes are preferable to hoppers, which can be wasteful, especially if it is decided to feed mash.

Water intended for chicks is best given in old-fashioned and cheap jam-jar drinkers; though small as they are, it is still possible for a chick to drown and so it is not a bad idea to place pebbles around the base of the drinker. Alternatively, and probably easier when changing the water twice daily, a thin piece of piping can be used. By placing the drinker on a piece of scrap hardboard, the water level is maintained and the shavings are less likely to get wet.

It used to be possible to buy galvanized grit hoppers that attach to the wall of the house, but I don't know if they are still available. If not, two troughs containing hard grit and oyster-shell can be included – though I know of several bantam keepers who sprinkle both types of grit over the ground to encourage their birds to forage.

Generally, there is a far greater choice of feeding and drinking vessels on the market than ever there was a couple of decades ago when all were galvanized and it seemed that 'Mr Eltex' had the monopoly. Nowadays it is possible to buy very efficient plastic utensils that are much cheaper. Their lifespan is not likely to be as long as the more traditional varieties, but at around a third of the price, they are well worth considering.

As an option to simply tying up any greens to the wire fence, a hanging flower basket could be given a new lease of life and used as a container for greenstuffs. Suspended at a height just out of normal reach of the birds, its contents will encourage them to jump and so provide exercise and occupation, which would be absent if the leaves were scattered on the floor.

CHAPTER 5

Breeding, Rearing and Hatching

Even if you have no intention of showing bantams, it is still important to breed any 'follow on' stock with the development of your own strain always foremost in your mind. It can be fun creating your own colour of a particular breed, and if you were to mix a pen of black Frizzle hens with a white cock bird, there is a good chance of the offspring being what is termed 'creole' in colour. A Frizzle cock put to a Rhode Island Red hen will produce a red Frizzle, whilst breeding black and white Rosecombs together will generally produce 'blue' chicks.

Tempting though it may be to experiment with a pen of cross-breeds in the hope of picking out a certain characteristic such as colour or feather type, it can be more difficult to guarantee success as one is never sure of their ancestry. A gene may lie dormant for several generations and then revert to type.

Serious fanciers breed not only for replacement stock but also to eradicate faults and

'Designer-bred' Creole Frizzle cock.

OPPOSITE: Suffolk Chequers male. Photo: Rupert Stephenson

improve on any points of the bird that may be weaker than required by the breed standards. Sometimes this is done by what is known as out-breeding, but as with all breeding 'programmes', one has to be careful not to introduce more problems than are being solved. As an example, the Frizzle's feathers can become too 'frizzly', and whilst most showmen merely introduce a smaller-feathered strain into their flock, others have been known to use a totally different breed such as the White Wyandotte. Although this mixture is normally successful in producing the perfect frizzled feather, it is not unknown for the resultant chicks to have the wrong type of comb, which will obviously not conform to the breed standard. Rumpless Game are often crossed with Old English to introduce fresh blood, but such a mating can produce youngsters with tails. Even if they themselves are perfect, there is always the possibility of faults emerging two or three hatchings later.

If you really want to set yourself a lifetime's task, there is no reason why you should not attempt to create a bantam strain from one of the breeds of large fowl that has still to be 'bantamized'. The most obvious way is to pick the smallest individuals you can find and breed from them, always mating with the smallest of the subsequent chicks. Try and buy in fresh blood at some stage to avoid the dangers of recessive in-breeding – though once your stock possesses all of the good points of the breed and none of the bad, there is a temptation to in-breed continually so as not to risk the introduction of faults. But fertility will undoubtedly be affected, and in my opinion it is a much better idea to line breed from a particularly good individual specimen from within the family line. Others will disagree, I am sure, and it makes sense to talk to a variety of experienced breeders before deciding one's own policy.

Depending on which breed(s) are eventually chosen, it may be sufficient to put a good cock bird with good hens and expect reasonable results. However, in some breeds or even types of breed, life is not so simple.

The Wyandotte types have always been a personal favourite, and the Whites with which I started my showing career hardly ever failed to win at either club or county level. They were an easy type to breed, hatching stock conforming to breed standards almost every time. Some fifteen years later I began breeding Black Wyandottes, but the results were not so easy to guarantee and I found that recessive, harmful genes periodically damaged any chances on the show bench by producing black legs rather than the yellow ones stipulated by the *Poultry Club Standards*. Tempted as I am by the beauty of Partridge Wyandottes, I have always shied away from them due to their even more complicated genetics and the need for separate pens of cock and hen breeders. The term is not, contrary to what some people believe, guaranteed to produce all cocks or all hens, but has more to do with their complicated plumage patterns. A hen-breeding cock must possess black/brown coloured breast feathers in order to produce well-marked females, whilst the black breast of a perfect show cockerel is derived from indistinctly marked cock-breeder hens.

So much for some of the theory behind successful breeding. The practicalities are just as fascinating, and must begin with choosing the right cock and hens.

SELECTING BREEDING STOCK

It has never been my intention to turn the keeping, breeding and showing of bantams into a mystical art form, and of course it is possible for the hobbyist with only a small pen of three or four birds at his or her disposal to be as successful as the more experienced bantam breeders. However, some general pointers regarding the ideal choice of breeding stock may help.

The cockerel is perhaps the most important part of the mating, and whilst every effort should be made to ensure that both cocks and hens are of the best possible quality, it is the cock bird that passes on the majority of plus points. Most of these desirable traits are inherited from his mother, and so it is useful to know the cock's parentage. Irrespective of the breed standards, the male should be well muscled and should 'feel' right when handled. Check the eyes, which should be bold and possess a round, black pupil. The legs must be strong in appearance, with the toes straight and not curled inwards. Check that the rear toes point backwards and not to one side (duck-footed). The comb should be as the breed standard indicates.

The hen needs to conform to the correct body size and shape of the breed, as these are perhaps the two main characteristics that she will pass down to the chicks.

Unlike commercial poultry-rearing units where large flocks of hens may be serviced by more than one cock, bantam breeders generally utilize separated pens containing one male to several females. You may think that by running two cock birds with one pen of hens, the chances of fertility will be doubly increased, but very often the opposite is true because they spend more time sparring with each other than they do in getting on with what they are there for. Also, it sometimes happens that both males concentrate on the same hens and ignore others. Even when a pen is made up of only one cockerel, it is still possible that a hen may remain infertile due to the simple fact that the male just doesn't 'fancy' her; I noticed this quite frequently when breeding English partridges.

The ideal number of hens to one cock varies from breed to breed, but it is traditional to buy and sell trios: that is, one cock to two hens. This is probably ideal for the more 'delicate' breeds of bantam, but some of the more vigorous types such as Rhode Islands will actually do better by having the number of hens per cock increased. A breeder of my acquaintance has a maxim of 'Definitely more than two, ideally a minimum of six', and claims that the optimum numbers are six to twelve per cock bird. Personally I think twelve is excessive, and would advocate six to eight as being the ideal. There is a school of thought that suggests cock birds of the heavier breeds cannot manage as many hens as those of the lighter varieties, but I have seen no evidence of this in my own pens.

If space and neighbours permit, it may be worth considering having a 'spare' breeding cock on hand. It should, of course, be selected with the same care as the original male, but it gives the fancier the opportunity to swap a 'tired' cock halfway through the season with a fresh and enthusiastic replacement.

Cocks can breed successfully for around five or six years, but their fertility decreases with age. Many breeders think it best to mate first-year cocks with second-year hens in order to achieve optimum fertility, virility, stamina and a good-sized egg which is ideal for hatching. Breeding pens should not be made up until the pullets are at point-of-lay stage, and the cockerels slightly older: around twenty-four weeks.

THE BREEDING SEASON

It is possible to breed chicks all the time the hens are laying, but to be sure of the best results, most breeders consider spring to be the ideal period. Cocks are at their most virile, eggs at their most fertile, and given good weather, the growing chicks will benefit from the summer months.

Any animals and fowl born or hatched in the spring will undoubtedly grow into bigger and stronger adults than those produced later in the year, but there is a school of thought that feels this to be undesirable in bantams, which, by their very nature, are supposed to be small. Breeding from the same strain every spring will increase the size noticeably after several generations. I offer this fact merely as 'food for thought', and doubt that any experienced poultry keeper would think it a valid enough reason to forgo spring breeding.

As the majority of small-scale bantam fanciers keep their flocks of birds penned as a breeding group all the year round, the eggs will be constantly fertile. New males should be introduced to the flock by late January, because although a cock bird will ejaculate enough semen into the hen's oviduct to fertilize a dozen or more eggs at one mating, sufficient time has to be allowed to ensure that mating has actually taken place. Serious egg laying should start around the end of February, and so it is reasonable to assume that they will definitely be fertile by March or April.

Extremely cold weather in February may well affect the fertility of both male and female birds. If it is intended to carry out serious breeding to ensure mature stock at the end of the summer, it is important to keep birds penned in as sheltered a place as possible, and perhaps rig up some temporary lighting in the shed from January onwards to give more 'daylight' hours and a warmer environment.

To obtain the best results, start feeding a breeder's ration at least six weeks before the first eggs for hatching are required. If, for one reason or another, it is impossible to source breeders' pellets or mash, then try and obtain a layers' ration containing a protein content of around 17 per cent. The addition of a little cod liver oil and a small amount of fishmeal (*see* the previous chapter) may also help in improving hatchability, especially during the early part of the season. Mixing it with the afternoon scratch feed of mixed corn is probably the easiest way to administer this, as the oil will help the fishmeal adhere to the grain.

When the pullets first begin laying, it is not unusual to come across irregular-shaped eggs. Only the best shaped and sized eggs should be considered for hatching, though don't worry about the occasional egg whose shell is coloured by blood, as this generally indicates that it has been produced by a young bird.

EGG COLLECTION AND STORAGE

During my years as a gamekeeper, I lost count of the number of times interested by-standers asked me how I kept the pheasants' eggs warm between collection and incubation! It is neither necessary nor desirable to keep fertile eggs warm. Indeed, a cool storage place of around 10°C (40°F) is ideal, as such a temperature will prolong the egg's life. I was always told that eggs should never be stored for longer than a week if being incubated by incubator, and no longer than ten days to a fortnight before being placed under a broody hen. The fresher the egg, the better chance it has of producing a strong healthy chick that hatches on time. Remember, however, that eggs set immediately after laying will probably not hatch because they have to be cooled first.

Collect eggs as many times daily as is practical, as this will help to keep the shell in good condition and prevent any hairline cracks occurring due to other birds using the same nest. A regular collection also prevents overheating in the warmer months. Mark the date it was laid on the shell with pencil so that you can be sure that the freshest eggs are set.

Only perfect eggs should be considered for hatching. Avoid including any that are misshapen, cracked, the wrong colour typical for the breed, or whose shells are chalky or lumpy in appearance or texture.

Once the ideal place for storage has been chosen, the eggs should be stored in egg trays, preferably on their sides rather than up-ended, and turned daily. When storing large quantities of game-bird eggs, I had a sheet of corrugated metal tilted at one end, into the channels of which I laid the eggs. Each morning I removed the bottom egg and placed it at the top of the 'incline', allowing all the others to gently roll down. This had the effect of turning all the eggs without the need to laboriously handle each one individually. It is unlikely that the amateur bantam keeper will be storing large numbers of eggs prior to incubation, but I include the system to illustrate my point.

If collected regularly from clean nest-boxes, there should be no need to wash the eggs, but

Any soiling on a hatching egg can be gently removed with tepid water and a pan scourer.

if, after a particularly wet spell, it does become necessary, they can be washed in warm (blood temperature) water. It is possible to purchase specifically manufactured cleaners and disinfectants, but the maker's instructions must be followed closely. The odd speck of dirt can be removed by gently scrubbing with a pan scourer. Immediately after washing, the eggs should be placed in a wire basket to cool off and dry before being stored in the usual way.

HATCHING UNDER BROODIES

I would always advocate hatching under broodies whenever possible, and I have a preference for the bantam over any of the larger fowl. However, provided that the breed is of the heavy, docile type, and will not rocket through the door every time someone comes close to the nest, there is no real reason why such a hen should not be used.

When a hen goes broody, she should be given a couple of days left alone in the chicken house before being moved to her new quarters where she will hatch and look after her chicks. If she is broody in the house from which one is collecting fertile eggs, make sure

she is not sitting on eggs that will be required for hatching at a later date, as this will affect their viability. Keep her sitting by including a couple of dummy eggs in the nest, or old eggs marked by a pencil so that they will not be inadvertently collected. When you try to put your hand underneath her, she should ruffle her feathers, sit tighter into the nest and attempt to give the back of your hand an almighty battering with her beak. Then and only then, can you consider her truly ready to begin her work.

A broody hen must have peace and quiet. Darkness or subdued lighting improves the steadiness of the hen. If a shed of some description is available, all that is required is a nest box of the right size, furnished with a wide board across the front so that she can turn around and be comfortable without risking any of the eggs disappearing over the side and becoming chilled. For this reason, it is not a bad idea to cut a grass turf, turn it upside-down, and place it in the bottom of the box. Alternatively, loose damp soil can be used and beaten into shape with the hand. It is advisable to wear a workman's glove, so that nothing hidden in the soil can cause injury. The nest should be made perfectly

A broody of this type is an ideal bird under which to hatch bantam chicks.

round, and the lowest point of the hollow should be the middle; not only does this help to prevent eggs from rolling out when the nest is built up with straw or wood-wool, but it also helps to retain moisture and humidity.

Probably the easiest way of dealing with a broody is by using the old-fashioned coop and run. The coop will have a sliding shutter, and a wire-netting run can be made as long as conveniently possible and must obviously fit flush with the front of the coop. It should also have some means of access.

Make a shallow, saucer-like depression in the ground, and fill it with clean straw or wood-wool before setting the coop snugly over the top. Fill any gaps around the outside with wood or bricks to prevent access by rats, and then include a couple of 'pot' eggs in the nest thus created. Broody hens should never be moved from the poultry house to the sitting box in daylight; the move must always be done at dusk, and the hen handled with great care.

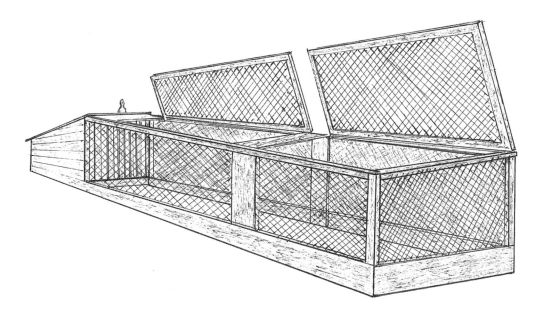

The old-fashioned coop and run makes an ideal unit for a broody bantam and her chicks. Care must be taken to ensure it is positioned on level ground, and any holes through which chicks might escape or rats enter must be securely blocked.

If, after twenty-four hours, she is sitting tightly and happily, the artificial eggs can be removed and replaced by fertile ones. The hen should also be encouraged to come out for a feed and a drink. A dust bath, kept dry under cover, will also be much appreciated. Once she is seen to be well settled and returns to her eggs of her own free will, it is possible to leave the front open so that she can come and go to the food and water as and when she pleases, saving you the trouble of lifting her on and off, and risking smashing one or two of the eggs in the process. Check that she is coming off the nest on a regular basis by looking for fresh droppings or diminishing food – some broodies are so intent on their responsibilities that they prefer not to leave the nest.

Feed only mixed grain, and never pellets or greens, as to do so will only encourage loose droppings and a greater chance of a soiled nest. If the nest does become dirty or any eggs broken, try and clear out the worst of the mess whilst the broody is out for her food and exercise. At the same time, wash off soiled eggs with aired water and dry them with kitchen roll.

Some situations where it is necessary to lift the broody on and off her nest twice daily cannot be avoided, and if this is the case, remember not to let her be away from the eggs for more than about ten minutes when she first begins sitting, up until about twenty minutes in the last week. Care must be taken in lifting the hen, and this is best done by placing a hand under each wing to make sure no eggs are lodged there; then take hold of the legs with the fingers and raise her gently, letting the wings rest on your wrists.

As the broody hen reaches the time when the chicks are due to hatch, she will probably not want to leave the eggs at all and will struggle to return to the nest. If hatching several batches of eggs with more than one broody, always ensure that the right hens return to the correct sitting box.

Do not be surprised if the broody has lost a great deal of weight by this stage. Although it might seem a peaceful experience sitting there turning a few eggs with her feet every now and then, the act of incubating has a detrimental effect on the hen. She will not only lose weight, but her plumage is likely to become noticeably dull and her comb and wattles pale. Some breast feathers will have fallen out completely and the bare patch exposed is known as the 'brooding spot', essential to transfer body heat to the eggs. As hatching time approaches, it will pay to spray aired water over the eggs or on to this brood patch, which will help with the all-important humidity, especially if experiencing a particularly dry and sunny spell of weather.

Just because a hen shows signs of being the perfect broody at the outset, it does not, unfortunately, follow that she will sit for the required length of time, and it is not unknown for a bird to leave the nest before her duties are complete. If you are lucky and the eggs are still warm and another broody hen is available, there is no reason why a successful hatch should not still take place; but otherwise the whole sorry event has to be put down to experience. Make sure that a diary of the daily happenings is kept, and make a note to remind oneself not to use this particular broody again unless as a last resort, or on eggs that are not crucial to the breeding programme.

Whilst on the subject of note keeping, most breeders agree that records are extremely useful. They can incorporate all manner of useful data, ranging from the date on which certain eggs were laid, through to what percentages were subsequently found to be infertile.

HATCHING WITH INCUBATORS

Basically, there are two types of incubator: cabinet machines and those that work on the 'still-air' principle. Unlike a cabinet machine, where the air is moved around the interior with the aid of a fan or rotating paddles, in the 'still-air' model the heat is supplied to the top of the machine. Because there is such a

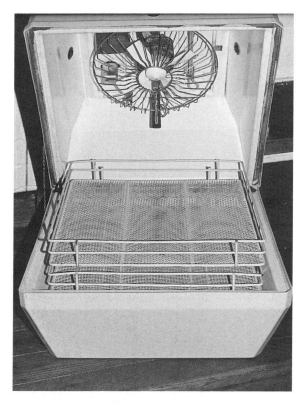

A small fan-assisted incubator. Photo: Gina Arnold

variance of temperature between the top and the bottom of the machine, it is vitally important to have the thermometer correctly situated above the egg tray, 6.4cm (2.5in) above the eggs, where the temperature should be 39°C (103°F).

Professional hatcheries use a combination of the two systems. The eggs are first placed in cabinet machines, which are now so sophisticated that, once set up according to the manufacturer's instructions, they require very little human intervention. Turning is done automatically, and the movement of a central rod up and down in order to tilt the trays, is operated by means of a time clock and limit switch. The fan or paddles control the very necessary air circulation, whilst humidity is checked and measured with a contact hygrometer or electronic sensors.

Two or three days before the eggs are due to hatch, they are generally transferred to a still-air incubator or a specially designed hatcher, where it is of vital importance that the correct humidity is maintained.

Cost and size will obviously preclude the use of large cabinet machines by the average bantam fancier, but there are several small tabletop machines available, which work on the same principle and should be more suitable. Use the manual supplied by the makers as a guide and then, with experience, you will be able to make small alterations and possibly improve subsequent hatching. Although the manual is only a very basic guide, it will advise the owner how to set up that particular machine.

The most common still-air models being produced today work in the same way as the old Glevum and Ironclad incubators. If these two types of still-air incubator were ever to be seen at a farm sale or in the advertisement columns of a newspaper or smallholder's magazine, I would strongly advise their purchase, as they produce some very successful hatches. Originally both were heated by paraffin, but over the years either the manufacturers or interested owners have converted them to electricity.

Heat is passed to the top of the incubator and the temperature regulated by means of a damper. This is raised or lowered by the expansion and contraction of an ether-filled capsule situated on a cradle near the top of the eggs. Humidity is supplied with the aid of two sliding water trays, which must be kept filled at all times, and at the base of the machine is an arrangement of two or three felts. These are removed one by one at intervals of a week and let out stale air.

Immediately below the egg tray is a flat tray on to which a piece of hessian is usually tacked. A glass panel fitted to the door situated at the front of the egg tray encourages the chicks, as they hatch and dry off, to make their way towards the light. They then fall through a gap and on to the hessian-covered

'Glevum'-type still-air
incubator showing its
various internal workings.
If an opportunity ever
arises to purchase a
similar-looking incubator,
do so!

Not much more than a polystyrene box, I am assured by its owner that this particular cabinet incubator gives excellent results.

tray, where they remain until the hatch is complete.

During the whole of the incubation period, the eggs in these small incubators need to be turned twice daily by hand, in order to prevent the developing embryo from sticking to the side of the shell. In a natural situation, it has been calculated that a broody hen turns her eggs once every thirty minutes or so. One way of ensuring that each egg has been turned is to mark them with a pencil cross on one side and a nought on the other. By making sure that all the crosses are showing in the morning, and the noughts at night, twice-daily turning seems to be sufficient to produce good chicks.

It is also important that any incubator is kept in a suitable shed or building, one that retains a constant temperature. Modern electric units are nowadays so compact and clean that it is tempting to situate them in the spare bedroom or study; but central heating and a stale airflow will affect the results. A brick building with good insulation is likely to produce a better hatch than a draughty wooden shed or, worse still, a tin one, where the temperature fluctuates dramatically every time the sun shines. The ideal temperature is probably around 16–21°C (60–70°F). Adequate ventilation is important as the machines take in fresh air and give back stale, and if the room feels at all stuffy it could be affecting the eggs. A concrete floor is a good idea – not only is it easier to clean and disinfect, but also, as hatching time approaches, it can be kept damp so as to improve humidity in the incubator.

Beware of the dangers of *too much* humidity. Possibly more eggs and potential chicks are lost in this way than for any other reason, and although many manuals recommend adding water to the incubator throughout incubation, in reality, very little is needed. The relative humidity should be around 60 per cent.

It is advisable to run the incubator empty for at least forty-eight hours so that any necessary temperature alterations can be made. Once the eggs have been set, do not adjust the thermometer for at least twelve hours as it can take quite some time for the centre of the egg to reach optimum incubation temperature. By allowing a brief 'settling in' period, it gives the incubator adequate time to correct itself, and it is to be hoped that no further adjustments will be necessary.

Remember to transfer eggs from the cool room a few hours before they are required, and to place them next to the incubator in order to bring them up to temperature. It is probably best to place the eggs in the incubator in the evening and to count the first day as being the one after they have been set.

Like this the eggs will have had time to reach the required incubation temperature.

CANDLING

At around two-and-a-half weeks of incubation, many fanciers like to candle their eggs and check for fertility. Some will also check at the end of the first week and look out for hairline cracks that may not have been obvious when the eggs were first selected.

White-shelled eggs are easier to examine and their contents are more obvious than eggs from breeds that lay brown or dark-coloured eggs, and for this reason it is perhaps advisable to delay the candling of these for a couple of days when the air sac and embryo development are more prominent.

Although it is sensible to remove any infertile eggs as they are obviously not going to hatch, the more eggs left in the incubator, the better it maintains temperature and humidity. Conversely, some poultry rearers believe that 'bad' eggs deprive the developing embryos of heat. It is, however, usual practice to remove both infertiles and dead-in-shells, the latter of which can be noticed by checking the development of the air sac when the egg is held up to a strong light. Candling machines are available commercially, but it is a simple matter to construct one's own: all you do is screw a light holder to the base of a wooden box from which a hole, roughly egg-shaped in size, has been cut in the top. By switching the light on and holding the egg over the hole, it is a simple matter to see whether or not the egg is fertile and at the correct stage of development. Easier still, cup a torch between your hands and hold the egg at the base of the 'tunnel' thus formed.

I was always told that the incubation period for bantams was nineteen days as opposed to the twenty-one days normally required by large fowl, but in reality, bantams can vary slightly depending on the freshness of the eggs and several other factors. You should, however, begin to see the first signs of chipping by the nineteenth day, and it is

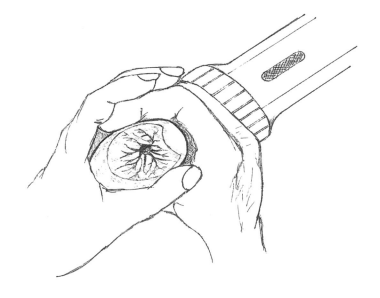

It is possible to check lighter-coloured eggs for fertility by 'candling' with the aid of an ordinary torch. Darker-shelled eggs may, however, require the construction of a special candling box.

One advantage of the Curfew 'see 'em hatch' incubator is, as the name suggests, that it is possible to check developments without lifting the lid and affecting the all-important humidity.

essential to have stopped turning the eggs a couple of days before this point is reached. From then on, there is no need to open the incubator again until the majority of chicks are dry and active – indeed, it is important to resist any temptation to do so, as it will undoubtedly affect hatchability.

When the majority have hatched and are dry, it is time to move them to their new home. Chicks still in their shells but cheeping will, I am afraid, probably be weak or deformed, and it is a kindness to open up the eggs and despatch the chicks as quickly and humanely as possible.

REARING UNDER BROODIES

Once it is obvious that there are no more chicks to hatch, all the empty shells should be removed and the broody given fresh food and water. She can be left to her own devices for another twenty-four hours or so as the chicks will live off the yolk sac absorbed during the last few hours of hatching.

When, for one reason or another, it is necessary to remove the hen and chicks from a sitting box to a new environment, now is the time to do it. If they were to remain 'in situ', I would recommend quietly and gently removing whatever material was chosen to make up the nest, and replacing it with fresh wood shavings. Place the hen and chicks in a cardboard box whilst this is being done, and they will come to no harm in the few minutes it takes.

If you want a broody to foster some newly hatched chicks, perhaps from an incubator, she should have been sitting for at least a week beforehand. At night, carefully remove an egg and replace it with a chick, gradually introducing them one at a time. Given a secure run, on short grass, safe from predators and positioned in a secluded place, with access to good quality chick crumbs and shallow water, it should be possible to let the broody get on with the job of rearing the chicks without too much intervention on your part.

It is better to keep the hen confined throughout the period of chicken rearing and not to allow her total liberty, otherwise you may well find that she takes her brood off for hikes into long wet grass, which is certainly not good for them.

ARTIFICIAL REARING

Small numbers of chicks can be reared quite successfully in a large cardboard box over which an ordinary light bulb has been fitted. I have seen the idea refined slightly by enclosing the bulb in a terracotta plant pot, around which the chicks can cluster and obtain warmth without getting burnt, but it is obviously necessary to ensure that the heat source is fitted in such a way that it does not set light to the cardboard. Infra-red lamps

Although used in this instance to move pheasant chicks, a properly constructed chick box helps in keeping newly hatched bantams safe and warm whilst being transported.

An infra-red light is probably the most popular source of heating amongst the majority of bantam breeders.

are easily available but are best fitted with dull emitters rather than the lamp usually used with other forms of livestock rearing. Calor gas is another option, and is certainly the choice of most gamekeepers, who believe it to produce the best chicks with the least number of problems. An electric hen brooder can also be extremely useful, but remember to check the underside for holes in the 'blanket' through which chicks have been known to squeeze and die.

Whatever method is chosen, it is important that the chicks learn the source of heat as soon as possible. When rearing is being carried out in a large shed, this is best achieved by making a circular surround of cardboard or hardboard. Not only are the birds guided towards the heat in this fashion, but there are also no corners in which they can crowd, suffocate or chill. After a few days, the surround can be moved away completely – though it is vitally important that a chick is never allowed to become chilled. Although placing a cold bird back under the heat seems to revive it, a high proportion of them develop gastro-intestinal problems or liver and kidney failure, from which they will never fully recover.

It is just as important not to let the chicks get too hot, and much can be learned by watching them closely for the first couple of hours – they are, after all, the best guides to the correct temperature. If they are seen to be huddled directly under the heater, they are undoubtedly too cold: there should be a central spot that is too hot, with a ring of comfortable chicks around it. If you watch an individual bird from this ring, you will notice that every so often it will get up, have a run around, try out the food and water and then return for a quick warm-up. If the circle of birds is right against the perimeter of the box or surround, then they are too hot and the space should be either enlarged or the heating arrangement raised.

Whilst a floor covering of wood shavings is ideal for bantam chicks with a mother to guide them, it may pay to make a floor covering of hessian for brooder-reared birds. This is only necessary for the first few days until the chicks learn that crumbs are more palatable than shavings. It is also a good idea

A surround, usually of cardboard, prevents chicks getting too far away from the heat source.

to use hessian under an electric hen, as it has been found that it will prevent birds from scraping a hole in the shavings and being unable to move out of the way when others crowd on top of them.

'HARDENING OFF'

Like plants in a greenhouse, bantams need hardening off. Reared by a broody hen, they will do this naturally, but artificially reared birds should have their heat source reduced from an initial temperature of about 32°C (90°F) until, by the third week, the temperature is down to around 21°C (70°F). By this time both naturally and artificially reared chicks should have access to an outside run in which, depending upon the weather, they will spend most of their day, only returning to 'mum' for a quick warm-up.

At three weeks, the bantam youngsters should, depending on the breed, start growing proper feathers on their backs and around the vent area. As the wing feathers develop more quickly than the rest, they are already gaining some protection against the elements until by the age of about six

weeks they are fully feathered and could, if required, be removed from the broody or heater. I would advocate encouraging the hardening off process for artificially reared chicks by turning off the heat source during the day from about three weeks of age – though of course much depends on weather conditions at the time.

By six weeks it should be possible to identify the sexes: the cockerels will have larger combs and a different stance to the pullets, and although they will not have an adult tail, its final shape will be defined. Truly adult plumage for both sexes begins to show itself at around twelve weeks, and the cockerels can crow quite loudly.

Cock birds surplus to requirements may need to be re-homed or killed if the neighbours are likely to complain. For purebred birds there may be the opportunity to sell them off at one of the seasonal Rare Breed sales, but for others there is probably no alternative but to cull them. The easiest and most humane method of killing any chicken is the traditional one of 'wringing' its neck. Try and find a fellow fancier who is prepared to show you exactly how to carry out this

There is sometimes no alternative but to cull young cockerels. Try and find an experienced person to demonstrate exactly how it is done before attempting it for the first time.

unpleasant but often necessary task before attempting it yourself. The expression, 'wringing' is actually a misnomer as the idea is to dislocate the neck Depending on whether you are left or right handed, this is best achieved by holding the bird's legs in one hand and taking the neck and hackle area in the other. By pulling down and twisting the head backwards simultaneously, death is instantaneous.

INTEGRATING STOCK

Integrating youngstock with old always has its problems, but the addition of an individual to an established pen may sometimes be necessary. Always introduce the newcomer to the perch at dusk when the residents have settled, and hopefully they will be so keen to carry on with their daily routine in the morning that they will be less inclined to spend time bullying the newcomer.

Even when a broody hen is returned to the pen from where she originated, there will still be the re-establishment of the pecking order. If she has brought up chicks that are to be housed with the flock, they can all be returned together when the youngsters are twelve weeks or so in age. Although there is safety in numbers, watch out for signs that one bird is being picked on more than the rest. If it becomes cowed and refuses to feed, there may be no alternative but to remove it from the pen altogether. Generally, however, a pecking order should soon be established and the newcomers accepted.

In an effort to keep a record on young and old (and in some breeds, individuals can be difficult to tell apart) it may pay to buy coloured leg rings from your local agricultural suppliers and 'colour code' birds of one year or of particular parentage. This will help with record keeping and future breeding programmes. Leg rings of different sizes are available, so make sure that the ones purchased are 'bantam'-sized. Some come in the form of spiralled plastic bands, and although they have the advantage of flexibility and can be applied or removed at will, they have the disadvantage of being more likely to be dislodged or broken. Always buy a couple more rings of each colour than you think you need, and immediately replace any that get lost or broken. Left until several bantams are ringless, it is difficult to remember who belongs to what. Closed bands are more secure and are usually attached when the birds are around twelve weeks of age. They are fitted by first placing the ring over the

Leg rings are a useful way of recognizing individual birds. Here a selection includes various sizes, colours, types and examples of those used in the Ringing Scheme.

three front toes and then gently lifting the single rear toe upwards so that the ring will slip over and on to the leg.

The Poultry Club runs a ringing scheme whereby bantam fanciers can purchase rings to aid in record keeping and security. Ringed birds are easily traced at all times, and even when sold, the ownership can be transferred through the Poultry Club's record system. There are other incidental advantages when considering membership of the scheme, not the least of which is the fact that the club provides prizes for ringed birds at a number of shows, including the National Championship and all championship shows.

CHAPTER 6

Showing and Preparation

When I first started keeping bantams back in the 1960s I had no intention of ever showing them, and it wasn't until a friend of my father's, a well-known show judge of the day, came and congratulated me on a particularly good batch of youngstock, that I gave it any consideration.

Showing your bantams adds an extra dimension to the hobby and is an incentive to continually improve your strain. Without the enticement of the show bench, it is easy for the bantam owner to slip into complacent ways, content to breed a clutch of replacement birds each year and be satisfied with the eggs and the enjoyment that a pen of bantams undoubtedly brings. Not that there is anything wrong with that philosophy, but exhibiting certainly extends the hobby and its pleasures.

There is the forward thinking necessary to ensure that particular birds are ready for a certain date and in peak condition; perusing

There is nothing wrong with just rearing a few chicks each year, but showing bantams brings its own pleasures.

OPPOSITE: Suffolk Chequers female. Photo: Rupert Stephenson

show schedules and filling in entry forms; and the last-minute preparations twenty-four hours' before. At the show venue, there is the nervous tension once the birds have been penned and you are banned from the hall until judging has been completed. Once exhibitors have been re-admitted, there is the excitement of dashing back to your birds to see whether or not they have been awarded a card. Of course one pretends to be casual about it, glancing at winning birds as you wander between the rows, but one eye is always straining for a glimpse of your pen. I doubt that this element of excitement ever leaves even the most experienced of exhibitors, and be they nine or ninety, the feeling is akin to those Christmas mornings as a child when one dashed downstairs to see what presents had been left under the tree!

Finally, just in case you still need some persuading, there is the added bonus of new friends and contacts made. Bantam keeping can be an insular hobby and attending shows will have many benefits, not least of which is the opportunity to meet breeders with whom you can exchange ideas and from whom you might be able to purchase new blood if and when it is felt necessary.

Unlike many other forms of livestock showing, there is very little backbiting and nastiness in the bantam world. Congratulations are sincerely meant, and criticism, which usually has to be asked for, is always constructive and made with the best interests of your birds and the breed in general at heart.

Showing is not, however, a simple matter of completing an entry form, rushing out to the

Showrooms are a good way of studying breed types and standards as well as providing the opportunity to meet up with fellow enthusiasts.

bantam house and grabbing a somewhat startled bird from its perch before stuffing it into a cardboard box and throwing it into a pen allocated by a show steward! It will pay the potential exhibitor to attend a few shows as a casual observer. Don't be nervous about asking even the simplest of questions for fear that you look a fool – everyone has to start somewhere, and it is a fairly safe bet that some of those picking up the prizes today knew less than you did when they first started.

A BRIEF HISTORY OF SHOWS

The period around the 1850s was an important one as regards British social history. Railways provided a fast and efficient way of travelling and transporting goods from place to place; photography was just coming into its own, making possible, amongst other things, a pictorial record; and for the first time there were equal numbers of people living in the country as there were in the towns. The better off who moved from the centres of cities into suburban areas had land at their disposal and turned it into gardens or 'allotments', where they could grow flowers and breed a few chickens. Victorians, especially the 'landed gentry', were fond of collections, particularly if the object in question was unusual and better than anything their neighbours might possess.

True bantams had been around for a while (notably Rosecombs and Sebrights), but now there was a surge of interest in miniaturizing some of the large breeds of fowl. Cock fighting had been abolished in 1849, and the breeding of gamecocks had been treated very seriously by both the humble rural villager who based his hatching methods on superstition, and by the more prosperous who attempted a more scientific and logical approach. Both now had birds they could not use, and although cock fighting continued illegally, the more law-abiding looked for a different way of showing off their stock.

Game cocks used for fighting had their combs 'dubbed', a practice that is now illegal. In addition, their tail feathers were trimmed and artificial spurs fixed to their legs immediately prior to a fight.

Some poultry breeders had been using exhibitions as a means of advertising their stock to the local community prior to 1850, and there is no doubt that this remained their main reason for showing until well after this date. As the history and lists of breeds outlined in Chapter 1 proves, the early 1900s saw the importation and further miniaturization of several American and many European breeds to the UK, the popularity of which increased as a result of being exhibited. Often, encouragement to show one's poultry at the local village flower and produce show was given by the local 'squire', who would provide both prizes and a venue. It was undoubtedly the combination of all these elements that kick-started poultry shows into something we recognize today.

BANTAM CLUBS AND SOCIETIES

Many popular breeds of bantam, and indeed large fowl, kept in the UK have their own breed club, formed specifically for the welfare and further development of their particular breed. Those that do not are, in several cases, looked after by the Rare Poultry Society.

The Poultry Club quite rightly prides itself on being 'the world's biggest poultry club'. It was founded in 1877, and exists to safeguard the interests of all pure and traditional breeds of poultry both in Britain and elsewhere. Not only does it play an important role in maintaining breed standards and safeguarding bloodlines, but it is also responsible for the annual National Championship Show at which over 6,000 birds are exhibited. Many of the various breed clubs hold a show in conjunction with this event, as well as club shows at a more local level.

Small fanciers' societies can be found everywhere, and in most cases membership is

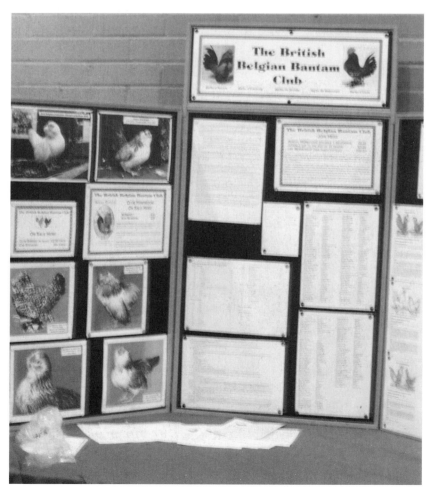

There are many clubs and societies, all with the interests of bantams at heart. Photo: Francesca Hobson

flourishing. There are numerous poultry clubs that have bantam enthusiasts amongst their members, and which hold bantam classes at their shows, but as far as I am aware, there are now only four clubs dealing exclusively with the bantam breeds and which have the word 'bantam' in their title. These are the Reading Bantam Club, the High Peak Bantam Club, the Cornish Bantam Fanciers and the West Essex Bantam Club. All of these are run in exactly the same way as any other 'normal' club, except that their shows only cater for bantams.

The majority of club shows are held in the autumn and winter months and can be at many and varied venues. I have shown my birds in places as diverse as village halls, working men's clubs, church committee rooms, above the bar rooms of pubs and, in the case of the Mirfield Fanciers' Society of the late 1960s, even a derelict house where 'hard-feather' was downstairs and 'soft-feather' classes were penned in the bedrooms. Summer shows tend to be much bigger affairs and held at agricultural shows, so it is sensible to gain experience at local level before moving on to greater things. The Poultry Club has a record of both the breed clubs' and general fanciers' societies, and will certainly do all they can to help the newcomer find either their nearest club or the secretary of a particular breed society. In fact, with access to the Internet, they can all be found listed on the Poultry Club's website.

SHOW OFFICIALS

Obviously the most important official is the judge or, rather, judges, because each section has a separate judge specializing in a

Judges have years of experience behind them and are usually only too happy to offer help and advice.

particular classification. Local show societies can, in theory, select whoever they like to judge, but if they intend their show to be recognized by the Poultry Club, the majority must come from the official list of qualified judges published annually. The list is divided into four categories: panels A, B, C and D. Judges work their way upwards in reverse order from D to A, usually over a period of several years, sometimes taking as long as ten to fifteen seasons before the top level is reached.

To qualify for inclusion, anyone who wishes to put him- or herself forward as a judge must eventually take several examinations; but to start with, most judges begin their career by judging their own breed types at local venues. The breed clubs will elect judges to officiate at their main club show, and once elected, an individual can be listed on panel D. Examinations to 'upgrade' to the next panel consist of both a written and practical section, and both parts of the exam are marked out of 100, with at least thirty points in each section being required for a pass.

The judging system is very complicated, and in some parts extremely technical. The Poultry Club can give further details and some pointers as to what is required of a 'would-be' judge. David Scrivener, an experienced and extremely knowledgeable breeder and judge, covers the subject in greater depth in his book *Exhibition Poultry Keeping* (also published by The Crowood Press).

It is obviously important that judges know the standards for each breed as set down by the Poultry Club, and which appear in *Poultry Standards* every ten years – the next one is due to be published in 2007. Equally, they should have some 'stock sense' and should not merely 'go by the book'. Unfortunately it appears that the two do not necessarily go hand in hand, and this comment has been made to me more than once whilst collating photos and information for this book. Judges are all volunteers and usually only receive the minimum of expenses for their troubles. Many do not even claim these, and stand the cost of travelling, meals and perhaps even an overnight stay, from their own pocket.

Stewards are usually experienced bantam keepers themselves and members of the committee; their role is to assist the judges. For instance, it is their job to ensure that competitors are allocated the right pen

Part of the steward's job includes ensuring that the right awards are displayed on the correct pen fronts! Photo: Francesca Hobson

number and have entered the correct classes. They are responsible for clearing the room before judging commences, and for ensuring that any entry cards or paperwork is placed face downwards so that the judges cannot possibly know the identity of any exhibitors. They also act as 'runners', marking down results and taking them to other committee members who have been allocated the tasks of noting down winners and writing out the prize cards – usually red for first, blue for second, yellow for third and green for fourth or 'highly commended'.

UNDERSTANDING THE SCHEDULE

It is obviously important to fully understand

Silver-pencilled Wyandottes are classified as 'soft-feathered'. Photo: Francesca Hobson

the schedule, and also to check out any clauses specific to the particular club or society. 'Must have been the property of the exhibitor and been in their possession for at least two months' or, 'Must have been bred by the exhibitor', may all find their way into the small print, and the comments must be observed and adhered to.

As a general rule, schedules are available from the show secretary at least a month in advance of the intended date. It is easy to let the enclosed entry form sit on the mantelpiece or on the desk, fully intending to 'deal with it later', but from my experience, it is far better to fill it in and return it as soon as possible to avoid the risk of applying for entry after the given closing date (usually about ten to fourteen days before).

In the first chapter of this book, most of the breeds mentioned indicate whether they are 'hard' or 'soft' feathered; 'heavy' or 'light' is included alongside their other characteristics. This should help in defining what classes are appropriate to your chosen breed. Briefly, they break down into:

Hard feather	including Old and Modern Game
Soft feather, heavy	including Wyandottes, Sussex, Orpingtons, Rhode Island Reds
Soft feather, light	all of the Mediterranean breeds, also including Polish and Silkies
True bantams	to include Pekins, Rosecombs, Sebrights, Japanese
Rare breeds	including Andalusian, Houdans, Booted (or Sabelpoot)

(Note: **Juvenile** classes refer to the exhibitor, and not the age of the bantam!)

AOV	All, or Any Other Variety not otherwise included on the schedule.
AOC	Any Other Colour, primarily intended for the less often seen colours of the more popular breeds.

Further details appertaining to the structure of shows at championship, regional or local level can be obtained from the Poultry Club website, *info@poultryclub.org*, or by writing to the secretary.

PREPARING BIRDS FOR THE SHOW

There are as many different ways of preparation as there are breeds of bantam. Some of the 'old hands' are quite secretive when it comes to handing out a few tricks of the trade to newcomers, but from my experience, the majority will be only too pleased to offer help and advice. A general timetable of preparation may be of use.

Some Weeks Before

Select possible show birds and begin conditioning them perhaps with the aid of a few special 'tit-bits', such as linseed, to their diet. If the standard of your particular bantam breed dictates that the legs should be yellow or orange in colour, the inclusion of maize to their feed will help enhance this coloration. Conversely, if your breed should have white legs, cut out any maize you may have got into the habit of including during the cold winter months. A favourite trick of game fowl exhibitors was to feed their birds from the top of a wire pen in order that they learned to stretch themselves and as a result, were encouraged into the 'reachy' position required. Old English Game should be fed a high quantity of mixed corn in order to retain the hard, tight feathering looked for by the judge.

Barred Wyandottes may find themselves in the 'AOC' class at some shows.

If your bantam stocks are large enough to allow such a measure, prepare a back-up bird; the unexpected can always happen no matter how careful you are, and it is useful to have a replacement ready and waiting on the 'substitutes' bench'.

By checking through the bantams thoroughly at this stage, there may be time enough to remove the odd soiled or broken feather and for a replacement to grow. This would certainly not be possible in cases when the bird is selected by torchlight on the eve of

105

the show. Trim up the beak and claws if necessary – I have found dog nail clippers perfect for the task, and they are cheaply and easily obtainable from your nearest pet suppliers.

Use these weeks to accustom the chosen birds to an exhibition pen. If space allows it and you have been able to build a penning shed as described in Chapter 3, the bantams can spend time in there and it will give you the ideal opportunity to handle the bird in a similar manner to the judge. Take the bird from its pen on occasions, and stand it on a firm surface. Obviously a small table would be ideal, but it should not be Formica-topped and slippery, otherwise the bantam could lose, rather than gain, confidence. Some gentle and judicious 'prodding' with a garden cane will help the bird accept the judging stick on the day, and should encourage it to show itself properly. If you can involve the rest of the family in this part of the training, so much the better: one of them could wear a white show coat and act the part of the judge. The more activity the bantam experiences, the better it will conduct itself on the day.

At this point, check and spray the bird for mites and lice. (More importantly, do not neglect this task on your return from the show.) Remember to send off your entry form, correctly completed and with a cheque for the entry fees.

Two or Three Days Before the Show

It will probably be necessary to give the birds a bath, but in the case of 'hard'-feathered stock, if it can be avoided, so much the better. Show experts vary in their opinion as to exactly just how many days beforehand washing should take place. I always found that forty-eight hours seemed the optimum for my White Wyandottes, allowing them time to dry but not to risk any soiling. Others suggest up to four days or a week ahead to ensure that the feathers are fully dry and will therefore 'hang' properly. This method also has the advantage of allowing feathers to regain their natural body oils.

Pure soap flakes are the best, but failing that, hair shampoo is a good 'second'. The water should be warm enough to allow the hands to be immersed, but obviously not so hot as to harm the bantam. Most exhibitors

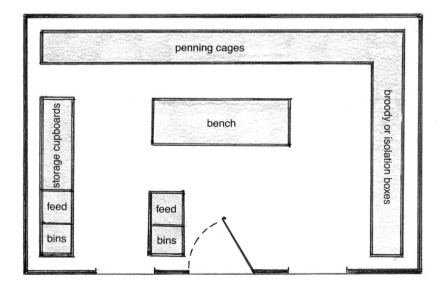

Typical layout of a penning shed.

Penning cages should be built at eye and waist level. It is possible to purchase either wire fronts or complete exhibition cages ready for use (the space under the cages is a useful storage area for carrying boxes).

tend towards three washes: the first containing soap, the second conditioner (or, in the case of white-feathered birds, a bag of good old-fashioned 'dolly-blue'), and a final rinse in tepid, clean water. Wash the legs and feet first, outside the bowl so that the worst of the dirt is excluded from the water. A nailbrush is ideal. Next, dunk the bird completely, obviously taking care to keep the head and eyes clear, and always cleaning in the direction of the feather growth.

Drying is best carried out by initially towel sponging the excess water before finishing off with a hair dryer fixed on a moderate setting. Attend to the neck and head area first and then, if it is a breed with heavy tail feathers, see to these next, forming them into perfect breed standards' shape. Lastly, see to the primary wing feathers, 'styling' them into the correct position. A night spent in a cardboard box close to a radiator or Rayburn in the

kitchen should ensure that the bird is looking perfect by morning. When completely dry, it can be returned to a clean show pen – but not before its legs and feet have been coated with either coconut oil or Vaseline.

Deeply serrated combs, for example rose combs found on the Wyandotte or Hamburg breeds, can be cleaned with the aid of a cotton bud and dried with kitchen paper. The old-timers of the show world used many special home-made preparations with which to enhance the comb; possible ingredients incorporated sherry, surgical spirit and after-shave. Other tips included the use of rouge, though most of today's fanciers seem content to use a drop of baby oil, carefully applied on the morning of the show.

Two Days Before the Show

It is not a bad idea to begin a course of water-soluble multi-vitamins and to continue the

Carrying boxes are a more secure and stable method of transporting birds than cardboard cartons.

treatment for three days afterwards. As explained elsewhere in this book, multi-vitamins lessen the possibilities of stress, and with it, the increased susceptibility to disease. Feed well *the night before*.

On the Morning

Be sure to allow sufficient travelling time to the venue, and double check the schedule in order to find out the latest times at which the bantams must be in their pens.

For many years, I felt very much the poor relation when I arrived at the show with my birds in supermarket cardboard boxes, when all those around me were transporting theirs in purpose-made wicker baskets. Eventually a favourite aunt had one of these made for me

at the local workshops for the blind, and I was able to carry my bantams with pride. In actual fact, it matters little what the carrying boxes are constructed of, but they should be large enough to allow the bantams sufficient room to turn around and to keep cool. Large boxes will also help in preventing feather damage.

Partitions should be included when transporting more than one bird: cock birds will have a scrap in the most unlikely of places, and it is embarrassing to arrive at a show only to find that your carefully prepared birds are splattered with blood.

Boxes made for the job have the advantage of being solid, but cardboard ones do not need periodic cleaning and take up no valuable

Feeders and drinkers fixed to the show pen will ensure that they do not get knocked over.

storage space as they can be disposed of immediately after use. No matter what sort is chosen, it goes without saying that they should be adequately ventilated.

At the Show

Don't panic, and do not let yourself feel hassled. See the stewards and obtain your pen number. Take time in completing those last-minute preparations with the oils and silk handkerchief before settling the birds into their allocated pen, their home for the next few hours.

Traditionally, birds are not usually fed or watered before judging, as a full crop could change the overall outline; however, once judging has finished, it should be a prime objective to ensure that the bantams are given water. At many venues the stewards will undertake this task, but it is as well to check that they have. I have never known a show society that did not provide drinking and feeding utensils (usually of the sort that are hooked to the pen sides and cannot be tipped over), but a couple carried 'just in case' take up very little room and should form part of your regular showing kit. Include some food as well – grain is better than pellets on this sort of occasion.

It is not always the case that competitors and the public are made to leave the hall before judging commences. Some, such as the massive National Championship Show held over two days each year at the N.A.C. Stoneleigh, in Warwickshire, actively encourage visitors in order that interested parties can observe and learn from the judges whilst these are 'at work'.

Normally you will not be allowed to remove your birds and take them home before a given time. Again, this could well be included in the rules and regulations found at the back of your schedule, but if not, ask a steward when you arrive in the morning.

After the Show

Get the bird(s) home as soon as possible and return it (them) to the penning shed together with food and water. It is a mistake to put it immediately with its mates in the main flock in case it has contracted some ailment from other poultry at the show. Whilst this is most unlikely, it is nevertheless a sensible precaution. Do not forget the spray or dusting powder to counter lice and fleas. Three days in 'isolation' should be sufficient to see that the bird has suffered no ill effects; by which time it will also have finished its course of multi-vitamins.

EGG COMPETITIONS

A less complicated way of becoming involved in the fun of showing with one's bantams is to exhibit their eggs. Although most egg competitions are run by poultry show societies as a minor part of their overall event, it is also sometimes possible to find egg classes at village shows, and at those run by horticultural societies. The general flower show often contains much non-horticultural produce, such as handicrafts, artwork, photography, cakes, honey, jams and home-made wine, and it is amongst this section that a class for 'plate of five eggs, white', may be found. A schedule is once again necessary to discover exactly what classes are available, and it will be no surprise to learn that even the humble egg has standards, and that a scale of points is looked for and awarded by experienced judges.

Each egg will be given marks for its uniformity, and in the case of a plate of eggs being shown, its similarity in size, shape and colour to its neighbours on the plate. Pullet eggs may be examined and awarded points differently to those laid by a mature hen: the base of the egg laid by the latter should be more pointed, but should also be neither too circular nor too narrow. Externally, the shell texture and its colour are also important, whilst internally, the yolk must be a golden yellow and free from blood streaks or spots. Its freshness will be indicated by the small air sac and the un-blemished surface of the yolk. A correct, healthy diet should

The egg classes are always interesting to see and can be a form of showing in their own right.
Photo: Francesca Hobson

ensure a good egg, but the addition of maize may well help in achieving a good colour of yolk.

It is possible to wash eggs in preparation for showing, but it is better if the eggs are kept clean naturally by using fresh straw in the nest boxes, and by changing the floor litter regularly so that dirt from the bantam's feet is removed before it reaches the nest box. Any odd specks of dirt on an otherwise clean egg can be gently removed with wire wool, though take care not to rub too hard and spoil the egg's natural 'bloom'.

At the larger poultry shows, these classes are well worth seeing, as the colours vary from the blue/green of the Araucana bantam's egg right through to the deep chocolate brown of the Marans and Welsummers.

THE RINGING SCHEME

Newcomers to the show world may read or hear of something called the 'ringing scheme', and wonder to what it refers. Basically, consecutively and individually numbered leg rings are purchased from the Poultry Club. Every year the rings are a different colour, and this enables everyone to see at a glance exactly when a bird has been hatched. As an added precaution, the year of issue is also noted on the ring itself, as some fanciers will 'recycle' rings from culled birds as a means of identifying youngstock. This is, however, only practical when used on breeding stock that will never be shown.

The scheme has been in existence for over a decade now, and the use of rings is proving extremely useful in a variety of ways, not least in that of security, as ringed birds can easily be identified if stolen. If and when a bird is sold on to another bantam fancier, it is a simple matter to inform the Poultry Club, who keeps a record of each ring registered by them.

I have visited many poultry shows in France and have noticed that all the birds exhibited there have been registered in a similar system, and that all are proud possessors of some very smart leg rings. Interestingly, at these French shows, not only is it forbidden to retrieve your birds until an allocated time, it is impossible, as the doors to the exhibition pens are secured by either a nylon or metal security tie that can only be cut by stewards equipped with pliers – a good way of preventing theft, as it is not unknown for a champion bird to have been 'spirited' away in the general mêlée and confusion at the end of a busy show.

The ringing scheme as set up by our own Poultry Club is also very useful in making bloodlines immediately recognizable; unlike other form of livestock, official pedigrees are not available for poultry, and this is one way of planning a scientific breeding programme. Different-sized rings are available for the various varieties of bantams, and the Ringing Scheme co-ordinator at the Poultry Club will be happy to offer help and advice.

THE JUDGE'S DECISION IS FINAL

Although a competitive streak and a desire to exhibit one's birds at their full potential are both necessary for success in the show tent, it should never be forgotten that showing is only a small part of the whole aspect of bantam keeping. Provided that your stock is of good quality, there is no reason why you should not come away from a show clutching a place card – if you are unlucky today, pick up another schedule and enter somewhere else in a few weeks' time. Over the years, you will lose some placings you should have won, and you will win some you should have lost! However, never lose your sense of perspective, or be seen to be a bad loser. The old competition rule, 'the judge's decision is final', applies just as much here as it does when entering for a chance to 'win a new kitchen' on the back of a cornflakes packet.

If you feel that you really have a grievance, mention it to a steward on the day, and if necessary follow it up with an official letter of complaint to the society committee; but never resort to a 'slanging' match in public. Ask the judges themselves why your bird was not placed, rather than speculate with other disgruntled exhibitors in a corner. With the courage of their convictions, most judges will be only too happy to tell you, and where possible, will offer constructive advice to ensure you stand a better chance next time.

FAKING IT

Throughout this chapter, I have attempted to give legitimate ways of improving one's chances, but it may be that you have unwittingly taken things a step too far and been guilty of what is known in the show world as 'faking it'.

'Faking it' is when measures have been taken in order to deliberately deceive the judges. Rinsing a white bird in water that has had a touch of 'dolly-blue' or a drop of glycerine added merely enhances the bird's existing beauty. Adding dyes or other false colouring to the washing water, as has been done by some unscrupulous exhibitors, is, however, literally a 'cover-up'. It is quite within the 'rules' to use oil on the legs of a bantam or an astringent on the comb to bring out the natural colour, but it is not acceptable to stain the legs with iodine or to mask a white spot on the lobe with a touch of lipstick. Fine lines, but nevertheless important ones.

That elusive red card will eventually be awarded, and is all the better received knowing it has been awarded to a bird, preferably of your own breeding, that is in natural peak condition as a result of all your hard work and effort.

CHAPTER 7

Health, Hygiene, Ailments and Diseases

Generally speaking, small flocks of carefully nurtured bantams should be strong and healthy enough to avoid the disease problems encountered by larger and more intensive breeders. Nevertheless, it would be foolish to assume that there is not much that can go wrong with poultry, and so do not attempt to cut corners in the daily routine. It is no coincidence that expressions such as 'cleanliness is next to godliness' and 'prevention is better than cure' have entered the everyday English language – they could have been coined with the poultry keeper in mind, and to abide by their dictums should be the fundamental aim of all bantam fanciers.

Thirty-odd years ago when I was forever pestering the 'old school' of stockmen for their advice on bantams, they all swore by the use of lime in their chicken runs, an annual application of creosote on their houses, 'Jeyes Fluid' or 'Lysol' to disinfect the floors or feeders, and Cooper's louse powder for the

Clean surroundings encourage healthy birds: in this case, a fine pen of Welsummers.

OPPOSITE: *Black Rosecomb male. Photo: Rupert Stephenson*

115

dark nooks and crannies of perch and nest box as well as for the birds themselves. Now, all but the lime have been withdrawn from public purchase, though fortunately there are plenty of other perhaps more scientific products on the market. Nevertheless, even some of these are being closely monitored and taken off the market as a result of European Directives. Confusingly, although creosote has been banned, there is on the market a new preservative called 'creosote', with none of the properties that the original product contained!

GENERAL HYGIENE MATTERS

Although we all start with the best of intentions, general hygiene can often slip when combining bantam keeping and a heavy work or family schedule. It is often the case that antiseptic footbaths, provided to prevent the accidental spreading of disease, become filled with rainwater and the concentration is never remixed, and so the effectiveness of such a trough of disinfectant or antiseptic is also diluted, probably by half, for subsequent users.

Naturally, the longer you keep and rear bantams on a certain piece of ground, the greater the chance of disease. Certain manufacturers claim that their products will kill microbes and parasites on open ground, but it is the opinion of most veterinary authorities that there is no known cheap and practical means of killing all infective soil-based bacteria. Where possible, move your stock to another area every few months or, at the very least, arrange two pens side by side so that one is periodically rested. An area that is growing grass is less likely to harbour disease than one that is a dust bowl in the summer and a mud bath in the winter.

If it has been necessary to cut grass in the 'resting' run before allowing access to your birds, do not leave it in a pile thinking that it will provide amusement for the bantams. It is more likely to become mouldy and infect them with one of the many fungal diseases.

Housing or rearing equipment that is only used seasonally will benefit from being cleaned and thoroughly disinfected. The disinfectant should be specifically approved for the poultry industry, and preferably 'Ministry approved'. 'Vanodine' is an excellent disinfectant as it can be used in the drinking water as well as for scrubbing down equipment or washing down the poultry house; the various strengths of mixing indicate its use at the respective dilutions. It is also a virucide, and listed by DEFRA as such. For those who aim to run their poultry unit organically, there are various products on the market, some of which are better than others. One that has been tested and proven to work is 'Eradicate', but it is only allowed to be classed as an inhibitor, purely because it would require the granting of a very expensive licence to claim that it actually 'eradicated'.

In the 'old days' it was considered a necessary chore to creosote sheds and houses every year, the general feeling being that a good coating not only protected the sheds from the elements but also controlled the spread of red mite. The great disadvantage of creosote, however, was that the shed had to be treated when it was out of use for a period of time, as the fumes were poisonous to most forms of livestock. Now that creosote is no longer available (its sale was made illegal in June 2003), there are many other wood-preserving products that can be used without endangering the health of any inhabitants. If the houses are kept clean and periodically disinfected with an appropriate product such as 'Poultry Shield' there is far less chance of encountering mites, lice and fleas, all of which can seriously affect the health, productivity and general well-being of the birds.

The need to provide and constantly replace soiled floor and nest-box litter has been emphasized in Chapter 3, but it is as well to remember that the dark confines of a nest box are an ideal environment for parasites and fungal spores. There can be no doubt that poultry appreciate the provision of straw in the laying area, and much of the pleasure of

Sheds and units should be cleaned and thoroughly disinfected at least once a year. Photo: Gina Arnold

Provided that straw is clean when baled and then stored in dry conditions, there should be little or no risk of fungal diseases when it is used as a nest or floor cover.

keeping bantams is derived from the simple task of collecting eggs from a bed of fresh golden straw. Provided that the straw was dry when baled and has not been subjected to damp in storage, there is no problem in using it, but it should be replaced regularly, and immediately if an egg is broken.

Discourage bantams from sleeping in the boxes overnight, as their droppings will encourage disease and parasites, and will soon soil the boxes as well as making the eggs dirty. A few nights spent removing birds from the nest-box and 'teaching' them to roost by physically placing them on the perches should overcome the problem; failing that, temporarily block off the openings to the boxes each evening, and this will soon discourage the practice.

STRESS

Stress can be a very real scenario when keeping, breeding and showing bantams. It is sometimes a subsidiary ailment caused by

Regular contact and handling lessens the risk of stress in bantams.

another situation, but it can also mask the signs of the real problem. Know your stock well enough to spot abnormal behaviour, however slight. Often the first sign of trouble is something you cannot really put your finger on – you just know the bird is not entirely happy. In a hobby of this nature it is a simple matter to identify individual bantams, and you will soon detect any changes in conversational tones, a slight withdrawal from the rest of the group, a droopy, hunched, fluffed appearance, or excessive thirst, lack of appetite, changes in droppings or loss of weight and condition. Trust your instincts, and worry if you notice that all is not well, even if there are no obvious symptoms and you have nothing more to go on than the look in a bird's eye or a hint of depression. It always pays to think 'cause and effect' when dealing with any form of livestock.

Until your stock becomes tame through careful and constant handling, every movement is likely to cause stress momentarily. Regular inspections are necessary for health reasons, and especially if you are considering showing; a trip to the show tent for the first time can be traumatic to the young bird, although the addition of multi-vitamins should help in reducing any problems. Some breeds of bantam are supposedly more susceptible to stress than others. Silkies are often cited as being a classic example, but I have never noticed this in any of my own stock, kept purely for their 'broody' potential.

More significant is the type of stress caused by bad management, either inadequate or too intensive, because this can provoke disease where it would otherwise be latent. Conversely, lice can be an annoyance to their host and therefore predispose the bird to stress, which in turn may lead to feather pecking or general bullying: 'cause and effect' is yet another dictum that should perhaps be engraved over the hen-house door.

LICE, MITES, FLEAS AND WORMS

Despite every effort towards keeping everything clean, bantams can still sometimes fall victim to minor infestations of lice, mites, worms and fleas. How seriously they are affected can vary, but all will need attention, and so it will pay the bantam keeper to routinely check birds for unwelcome 'guests'. Birds that are handled regularly in preparation for showing are less likely to suffer than 'garden' pets because they are under more constant observation, but all could benefit from a weekly 'nit-picking' session!

Lice

When large numbers of lice are active on a bird, the most obvious signs are greyish-white eggs around the base of the vent feathers. In really bad cases they will also be seen around the 'armpit' area under the wings. The lice themselves are about the size

of a pinhead and light brown in colour. Infestations are usually worse in the spring and summer months, causing irritation to adults and sometimes severe anaemia in young birds. Fortunately they are one of the easiest parasites to get rid of and, once discovered, can be eradicated by a proprietary louse powder purchased from your veterinary surgeon or local agricultural suppliers.

An alternative would be to use a spray. Some sprays are sold as being suitable only for pigeons or aviary birds and are therefore not licensed for sale as a poultry product, though in actual fact, most contain exactly the same ingredients as in products sold for chickens or bantams; if you are lucky enough to have an experienced livestock breeder as your supplier, he or she will probably be happy to advise you in your choice. (Be aware, however, of contravening the laws set out by the Veterinary Medicines Directorate.) Follow the instructions from them as well as those on the shaker or spray, and take particular notice as to when another application is required in order to kill off lice that will have hatched since the initial treatment. 'Poultry Shield' is very useful here as it can be sprayed directly on to the bantams and will effectively kill lice and northern mite, including their eggs.

Mites

Some mite infestations are more serious than others, but all types of mite are more likely to be a problem than fleas or lice. They all have a similar life cycle, a short one of seven to eleven days. Traditional stand-bys included the use of Stockholm tar applied well in among the feather roots, but today there are several medications on the market that are much more effective.

Northern Mite
These insects can cause a huge amount of 'damage' before being noticed. Similar in size to the red mite, they are normally grey to black in colour, but can show as red if they have just gorged on blood. We are given to

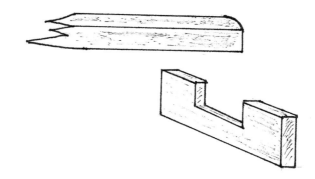

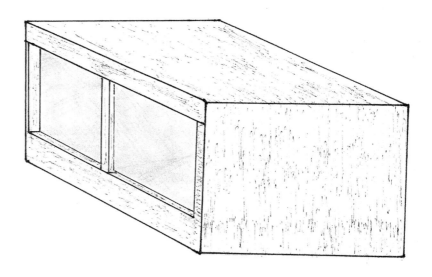

Dark corners of the shed, in nest boxes and around perch attachments require regular and careful checking for any signs of mite.

believe that, as with lice, the mites' preferred home is around the vent area (which can look slightly damp when affected); however, David Bland, a knowledgeable and extremely experienced poultry keeper of many decades, tells me that he has noticed them most frequently around the head and also under the wings. According to David, birds suffering from a heavy infestation tend also to become scabby on the comb, face and wattles. The preferred treatment is a spray containing pyrethrum,

but whatever product is chosen, northern mite is notoriously difficult to eradicate and repeated spraying will probably be necessary.

Red Mite

Unlike northern mite, which live on their host all the time, red mite live and breed in crevices found in the house and only 'hop' on to the bantam for a quick meal. It is pointless looking for them on the birds: they will be more readily seen by inspecting the perches and nest boxes. Often a substance the colour and texture of cigarette ash will be noticed: this is the mites' faeces. Red mite can live for many months without feeding, and have been known to lie dormant for several years. They may arrive via egg trays, contaminated feed or new stock, or wild birds. (Wild birds can be a problem in carrying both parasites and disease to domestic fowl; some breeders recommend feeding inside so that wild birds cannot defecate on the feed and possibly transmit serious diseases such as avian tuberculosis.)

Despite their name, red mites are also grey in colour and only appear red after feeding. Possibly the easiest way to differentiate between red and northern mite is to notice where they are found, but even though the red might be the lesser of two evils, both require eradicating as soon as is practicable.

Depluming (Feather) Mites

These mites live at the base of the feathers and attack the feather shafts. In really severe cases they can cause the feathers actually to fall out, or may irritate the bird so badly that it begins to pull out its own feathers. The symptoms are sometimes mistaken for feather pecking, especially as it occurs around the head and neck. Some long-feathered varieties of bantam will characteristically pull each other's feathers out during the breeding season, a trait rather romantically known as 'the lover's kiss', but this is not to be confused with the more mundane problems of the feather (depluming) mite. Difficult to eradicate, it is best treated by spraying as for northern and red mite.

An alternative would be to use 'Eprinex' which, although not licensed for poultry, works well. For a bantam, the rate of application would be four drops on the skin. (It is important to note that it is illegal to use products on animals or birds for which they are not specifically intended unless there is no alternative, in which case it must be prescribed by a qualified vet.) It has to be said

that depluming mite are very rare under good hygiene conditions.

Scaly Leg Mite

I was given my first-ever White Wyandotte bantam cock at the age of thirteen, and went on to breed several show winners as a result of its brilliant bloodline. Unfortunately, it also came complete with a rather bad case of scaly leg mite, and instructions to 'paint its legs with paraffin'.

Despite being supposedly easy to kill, I found it impossible to get rid of the problem by this method, and eventually the scales began to lift and peel. I now know that this was caused by the mite burrowing under the scales of the leg and breeding in the cavities thus created, but at the time I became desperate enough to try all the remedies offered by well meaning poultrymen. These included engine oil and surgical spirit, none of which could have been very comfortable for the poor old cock bird. Surgical spirit in particular, if applied at the end of the day, can burn the skin as the legs come into contact with the breast whilst the 'patient' is roosting. Sulphur ointment was also recommended.

However, subsequent bantams suffering from the same ailment have fared much better, as I have discovered that the product 'Vaseline' not only blocks off the supply of oxygen to the scaly leg mite, but also helps to loosen the old scales and strengthen the new as they grow. Applied in a seven-day cycle, it kills and heals at the same time.

Some years ago I had a horse that suffered from a skin affliction known as 'sweet-itch', and although there were certain preparations on the market, they were very expensive. Then one day someone suggested that I use benzyl benzoate, which could be bought at a fraction of the price from any chemist. Apparently it is also used for a very unpleasant human ailment, and I began to enjoy seeing other customer's reactions whenever I asked for the product! Visiting a new pharmacist one day, she asked me if I wanted it to treat scaly leg in chickens – up until then

Some parasitic worms can re-infect by passing through the bird, being taken in by earthworms and then eaten by the bantam, thus causing a continual cycle.

I hadn't even considered its application in such a situation, but subsequently I have found that it is by far the preferred treatment of most poultry fanciers.

Feather-legged types are particularly susceptible to scaly leg, and as it is more difficult to spot in amongst the feathers, owners of such breeds need to be more vigilant. Remember that the types of mite that live permanently on the birds can easily transfer from one bantam to another, or from a broody to her chicks.

Fleas

Poultry fleas tend to live in the dust of nest boxes and floors. The female fleas often lay their eggs in dirty nest-box litter; these hatch in a week and mature in a month. Although they do not actually live on the birds, they irritate the bantams by biting. The birds can be dusted, but will quickly become re-infected unless the house is cleaned and disinfectant used.

Hen fleas have a slightly different lifestyle and live in the downy parts of the bird's feather. They are orange in colour, and are sometimes confused with red mite.

Worms

Parasitic worms are generally referred to as helminths – hence the treatments for worms are termed anthelmintics. Some helminths can live inside a host and cause no harm; others, however, can lead to serious problems. The secret in dealing with them is to break the cycle by which the birds become infected.

Basically, fertilized worm eggs are excreted and are either directly picked up by other birds, or become larvae in some other species first – in the earthworm, for example, or perhaps in tiny snails that are later eaten by the bantam. Youngstock should be first wormed at about sixteen weeks of age, and regularly thereafter. Just how regular the follow-up treatment needs to be depends on several factors: every two to three months if the ground is continually stocked, but perhaps only once or twice a year if it is rested. September is normally a good time, as it is the end of the season and birds are going into the moult. Always worm just before you switch birds to a new pen, and those newly purchased show bantams should also be wormed as a precautionary measure before they leave the isolation unit.

When I was keepering, I used to have 'Flubenvet' mixed into the pheasant food. At the time I was rearing literally thousands of young pheasants and so there was no problem in getting the manufacturers to do this, but they obviously will not mix small quantities for the hobbyist with only a few birds. Fortunately, it is possible to obtain Flubenvet in more manageable sizes, and add it to the food yourself. If your general agricultural suppliers do not stock it, the local game farm should be able to assist.

The management of worm-infected land is similar to that used when attempting to prevent any heavy parasitic build-up: run young birds only on fresh land that has not been previously stocked by adults (who are better able to withstand worms) for at least six months and preferably a whole year. It will not necessarily clear up the problem altogether, but it will certainly help.

Roundworms
Roundworms are, as their name suggests, round and smooth, whereas the tapeworm is segmented. The female roundworm lays her eggs in the intestines of domesticated fowl; they are eventually excreted before being picked up by other birds, hatching in the intestine of the new host; and the whole process begins again.

Tapeworms
Tapeworm eggs may also pass through the bird via its faeces, or be retained in one of the segmented sections of the worm. These sections periodically break off and are also excreted, at which point they are sometimes eaten by small invertebrates who become the temporary host. If a bantam then eats that host, the cycle continues, but unlike the roundworm, if only the eggs or segmented sections are ingested by the bird, they will not develop into more tapeworms.

Gapeworms
Gapeworms are not often seen in poultry, and are more commonly found in the shooting world. They are worth a mention as it is possible for bantams to become infected from ground where pheasants and other game have had access. The female gapeworm lays her eggs in the windpipe, which are then coughed up and swallowed by the host, eventually being excreted in its droppings, where the larvae develops before being either picked up by other birds or ingested by earthworms. They can live for several years in an earthworm, and if the host is eaten by a bird the larvae find their way into the bird's lungs and, finally, the windpipe as adults to continue the cycle. Infected pheasants 'snick' or sneeze, but poultry tend to 'reach for the stars' so as to straighten their necks and thus assist breathing.

Hairworms
Probably the least known of all worm infestations, the hairworm can, nevertheless, cause problems to the poultry keeper. Rarely encountered except on overstocked land, they can, nevertheless, stress the birds to such an extent that Mareks may follow as a secondary disease. Unlike other worms they are not easy to see, even by experienced poultry breeders. They tend to be found in the 'blind gut' (caeca) of the bird.

Indications as to the presence of any worms vary from the telltale breathing difficulties as a result of the gapeworm (not to be confused with a cold, which is caused by a virus) to the unfortunate death of birds left untreated for hairworms. Generally, symptoms include a loss of weight, even though the amount of food being consumed is maintained; diarrhoea; pink, rather than red, combs; and an unusual lethargy. Good pen management and regular worming with 'Flubenvet' will, however, go a long way towards ensuring that worms never become a serious problem.

SOME COMMON AILMENTS

Before moving on to specific diseases, it is worth outlining some potential problems, but it should always be remembered that many bantam fanciers go through a whole lifetime of poultry keeping without ever encountering a single ailment. 'Forewarned' is, however, 'forearmed'.

Bumble Foot

Sometimes swellings can be noticed on a bantam's foot; generally known as 'bumble foot', it can have three causes. Firstly, a hot swelling on the foot pad and between the toes can be caused by a splinter which, if it can be seen and removed, should cause no further problems. Bathing the foot with warm water and an antiseptic such as TCP will aid the healing process. Swollen pads noticed under each feet can be the result of the birds having to jump down from too high a perch on to hard ground. Normally it is the heavier breed types that are affected, and the obvious answer is to lower the perches; but if for any reason this is impossible (for example, where bantams choose to roost in the rafters and beams of an old barn), make sure there is a soft landing area.

The third cause will often show the same symptoms as the first, but the swelling may advance further up the leg and is due to a staphylococci infection (subdermal abscess), best diagnosed by a veterinary surgeon. In addition to the outward signs, the foot and ankle are sometimes hot to the touch, but the problem can be rectified by the use of a broad-spectrum antibiotic. It is important that the vet makes the correct diagnosis as some have been known to lance the foot in order to remove 'pus' which is not there.

Colds

It is possible for bantams to suffer from colds usually caused by a virus. Symptoms include sneezing and watery eyes and nostrils – although these could also be brought about by other problems, such as being kept in a run full of long, seeding grass: yes, bantams can suffer from hay fever. Another cause could be the presence of ammonia in the poultry house due to wet, dirty conditions or poor ventilation. If you can smell ammonia in the house, then there is a problem that needs immediate attention.

Dry, dusty litter may also be the cause of cold symptoms. I remember some of the old bantam fanciers in Yorkshire using peat as a floor covering, and in the summer months they made it a daily practice to damp down the litter with a watering can.

All these forms of sneezing can be got over by altering and improving the environment. No medicated treatment is necessary, although a five-day period of multi-vitamins in the water will undoubtedly assist the birds in getting over their problems.

True respiratory infections are best cured by a week's course of 'Erythrocin' in their water. Available only from a veterinary surgeon, it is a very useful antibiotic. Not only will it combat any respiratory disease but also many staphylococci infections. 'Aureomycin' used to be the preferred drug, but is now somewhat dated as far as colds are concerned. Should the problem of mycoplasma occur then 'Tylan 200' is the answer.

Crop Binding

Not to be confused with sour crop, congestion of the crop is quite common in all forms of poultry. The crop will be noticeable even if the

Poor night-time ventilation can cause severe viral infections.

bird has not recently fed, and is solid to the touch. It is usually caused by some form of obstruction, quite often debris such as litter, long grass, or even discarded feathers, that prevents food from passing through and into the gizzard. Professional poultry keepers may well suggest despatching the bird, but with a bantam 'pet' or potential rosette winner, it is worth seeing if your vet is prepared to operate.

Sour Crop

Sour crop is normally caused by a fungal infection. The main difference between the two is that the crop is soft to the touch and the bird's breath will smell sour. Home treatment is possible: try to free the problem by holding the bird upside down and gently massage or squeeze the crop with your hand.

If successful, the bird is best isolated for a day or two. 'Vanodine' in the drinking water will kill off the fungus blocking the base of the crop, and should prevent re-infection.

Egg Binding

Egg binding is normally found either in young hens when they first commence laying, or it can sometimes be caused by an egg being broken in the passage of the oviduct. I well remember the astonished expression on a relative's face when they visited unan-nounced and found me holding the rear end of a White Wyandotte pullet over the spout of a recently boiled kettle! There was logic to my madness, as the steam from hot water will often loosen any internal fat and expand the bantam's pelvic area.

Egg Eating

This complaint can often occur as a result of an egg becoming broken in the nest box, and the resultant mess being pecked at out of curiosity. Another reason could be the absence of sufficient oyster-shell grit, causing the hens to lay thin-shelled eggs which are therefore more easily broken. A clean, dark nest-box keeps the eggs out of sight and therefore less tempting. A regular supply of both oyster shell and mixed grit, and the inclusion of fresh greens or even 'Peck-a-Blocks' in the house will help in preventing the problem which, once it has started, is very difficult to stop. Blown eggs filled with strong mustard may be unpleasant enough to prevent a particular bird from being tempted again, but as chickens do not actually have a good sense of taste, this well known remedy is unlikely to work.

Egg Peritonitis

It occasionally happens that a laying bird secretes an egg yolk into an alien part of its body, causing an inflammation of the peritoneum. David Bland, in his book *Practical Poultry Keeping* (The Crowood Press), describes the symptoms thus:

> The first sign that something is wrong is when the bird stays in the house and stands around the nest-box. Her comb becomes purple and the resultant septicaemia produces a very high body temperature. She will not feed but if strong enough, will perch that night. The following morning she will be found dead under the perch, lying on her side with her neck stretched and curved, the beak pointing to the crop, the comb purple. The bird dies in agony, and this is why, once the symptoms have been clarified, it should be put down immediately to avoid further unnecessary suffering.

Feather Pecking

Feather pecking is generally associated with overcrowding and is, like egg eating, a problem sometimes difficult to eradicate.

When pecked constantly, the feather follicle becomes damaged and will not function sufficiently well enough to produce more feathers. This would obviously end any hope of successfully showing the bird. It has been suggested that feather pecking can also be triggered as a result of a mineral deficiency in the bird's diet, but that given a balanced poultry feed and a plentiful supply of greens, it is unlikely to be more than boredom. (According to some experienced poultry keepers, feather pecking can also occur as a result of feeding pellets rather than dry mash, although I personally do not subscribe to the theory.)

Whatever the cause, a bigger run, dust baths, and a constant supply of vegetables may help them forget the habit. Alternatively, 'bits' can be fitted: these are a semi-circular, rounded piece of plastic fitting between the upper and lower mandibles and held in place by the nostril spaces. Obviously this does not prevent feeding, merely the beak closing on a feather. The application of a good anti-peck spray will also help. Bantams moult their old feathers and produce new ones each autumn, during which time they will look out of condition. This is a natural process, and nothing to do with the problem of feather pecking. It is also worth checking the bird's feathers in case there is an infestation of mites or lice.

Gizzard Impaction

As has been shown in Chapter 4, it is important that all poultry has a regular supply of hard grit with which to digest their food once it arrives in the gizzard. Week-old pheasant poults in my charge used to suffer periodically from an impacted gizzard due to their consuming wood shavings and not having the necessary 'grinding' materials in their gizzard to cope. Bantams are unlikely to suffer, which is perhaps just as well, because to diagnose the problem efficiently, it is necessary to cut open the gizzard!

Lameness

There are a number of possible reasons for lameness, ranging from the leg of the bird

Small pens may encourage feather pecking, but covered ones such as this are useful in preventing 'brassiness' in light-coloured show birds.

becoming injured as it dropped wrongly from the perch or pop-hole ramp, to it being a symptom of a potentially serious disease such as Marek's. Assuming it to be a 'one-off' problem, it may be a result of either insufficient or an unbalanced mineral content in the food, which, over a period of time, could cause damage to the ligaments. The addition of multi-vitamins to the drinking water will do no harm, but it is best not to mix with any other drugs or medications without first speaking to a vet. If a bird is seen constantly squatting rather than merely appearing lame, it is possibly suffering from cramp. Noticed mainly in growing stock, cramp is usually caused by either poor circulation or

poor conditions. Changing the conditions and housing affected birds intensively for a while, should effect a full recovery. A bantam that has strained a ligament can take many weeks to recover.

Overgrown Nails and Beaks

Kept out of doors and with plenty of scratching material available, it should be unnecessary to trim the toenails. Occasionally, however, one nail can become damaged, and it is a simple matter to rectify the situation with a sharp pair of dog or human nail clippers. By holding the foot up to the light it should be possible to see where the blood vessels or 'quick' ends and the actual nail begins. It is important not to cut into the 'quick' otherwise it will bleed quite badly, and will require cauterizing with a hot blade. Sometimes it will be necessary to trim up an overgrown beak, but only the top is trimmed until it meets comfortably with the lower mandible.

Poisons

In the interests of health, hygiene and keeping the neighbours happy, it will be necessary to keep the bantam yard free from rats and mice. It is perhaps obvious to say that rat poisons should not be placed close to anywhere the bantams have access, but beware of the possibility of the rats themselves carrying poison into the runs. Other unlikely 'poisoning' can come from the well intentioned feeding of fruit cake: I once wiped out a pen of potentially show-winning Black Wyandottes by doing just that. The sudden and unexplicable death of young, otherwise healthy stock, led me to have a post mortem carried out, and the results showed that the stale fruit had fermented in the digestive system. Of a similar nature, poisoning can be produced by a toxin originating from a fungus found in foodstuffs such as nuts, sunflower seeds and grain that have been stored in warm, humid conditions. Known as aflatoxicosis, symptoms can include lack of appetite, general lethargy, haemorrhage, and liver and kidney failure. Surprisingly there appears to be no such danger in the inclusion of a yew tree in or around the poultry yard, even though it is well known that the yew is deadly poisonous. The only part that is not is the flesh of the fruit. The seeds within are, and it is thought that the only reason that wild birds and poultry remain unaffected is because the seeds pass through the body untouched.

Predation

Bantams, especially young stock, can provide an easy meal for predators such as rats, stoats, magpies and even tawny owls and sparrow-hawks, who are surprisingly adept at entering buildings and runs through even quite small gaps. There is very little one can do about the latter two species, except to deter them by means of silver foil strung across the pen on cotton, but it may pay to have poison laid for rats and Larsen traps set for magpies. Not only will the successful capture of these birds protect chicks and eggs, but it will also help in preserving the local songbird population. Traps must be checked twice daily, and it should become part of the daily routine to do so.

It is always a heart-stopping moment to enter the bantam run and see birds lying dead. The work of a fox is easy to identify: birds killed by stoats or owls will have puncture marks and will be eaten from the base of the head, and rats generally kill and eat from the throat. Small chicks may be removed completely.

Prolapse

A prolapse is usually seen when a young pullet lays too large an egg, or in an old hen whose muscles have relaxed too much. It occurs when the vent muscles are pushed out with an egg still inside, causing a proportion of the organs to be visible. If the egg can be seen, it should be gently removed, breaking it if necessary, and then clean up all protruding parts before attempting to reinstate them with the aid of a lubricating jelly. Isolate the hen in a quiet place for a few days, and if the problem re-occurs, there is no alternative but to repeat the procedure. As amateurish as the method sounds, from experience there is a good chance of success, although great care must always be taken not to allow the prolapse to become infected. Apart from lessening the stress on the affected bird by placing it in an isolated box, if it were to be returned immediately to the flock and the intestines became exposed for a second time, there is a good chance that it may encourage the others to peck out of curiosity and maybe induce a spate of vent pecking which is as difficult to eradicate as feather pecking or egg eating.

Although a rare occurrence, it is, however, not unknown for bantam chicks to be taken by Tawny Owls. Always ensure that birds are secure from any predators at night. Photo: Gina Arnold

DISEASE

Apart from the parasites already mentioned, bacteria, viruses and fungi cause the main diseases. The list of potential diseases is a long one, but only a few are likely to be encountered by the small-scale bantam fancier, and so I have only included a few of the most common. Similarly, the information given can only ever hope to give but a rough idea as to what to look out for and how to deal with it. The main thing is not to panic and immediately assume the worst: if in doubt ask the advice of a neighbouring fancier or your local veterinary surgeon, who may or may not have a wide experience in poultry-related matters, depending on whether the practice is town- or country-based.

Aspergillosis

This is a respiratory problem usually resulting from inhaling fungal spores generated in damp or mouldy conditions often found in badly kept hay or straw. Damp shavings, especially those purchased in bale form rather than loose in bags, are another possible source. (When research was carried out in the broiler industry a few years ago, it was discovered that 80 per cent of the bales tested contained harmful spores.) Chicks are particularly susceptible, and the spores can penetrate the shell of an incubating egg. Symptoms can include laboured breathing, lethargy and increased thirst. Sometimes birds can be unsteady on their legs, maybe even walking backwards with their head to one side. It is better to practise avoidance by means of good hygiene, as there is no effective treatment.

Coccidiosis

All land is affected with the oocysts of coccidiosis, and if it is not properly rested, will undoubtedly have a high count, which could possibly infect young birds reared there. Chicks reared under a broody hen will, however, have ample immunity.

Coccidiosis is species specific, meaning that although a bantam and a pheasant may have access to the same ground, it is impossible for the bantam to affect the pheasant, and vice versa (though obviously the bantam or pheasant could infect another of its kind). It normally affects young birds between perhaps three and ten weeks of age. There are no specific symptoms, but the affected bird soon becomes noticeable by its dejected appearance, ruffled feathers and occasionally blood-stained droppings. The droppings, whether blood-flecked or not, will probably be white, sticky and copious. A lack of interest in food is another indicator.

There are many forms of this disease, the one that affects the bird's caecum or 'blind gut' being the most common. Usually all are present, but the main give-away is a mass of cheese-like substance, apparent when the carcass is opened up. The disease is caused by a microscopic parasite that ruptures the blood vessels in the intestines; this, in turn, infects millions of cell vessels, and the cheesy material is in fact a mass of these cells, which have died. In caecal coccidiosis, the 'blind gut' will be found to be full of clotted blood. Blood vessels may also be inflamed, depending upon which form of the disease is most virulent.

Most poultry foods contain an added coccidiostat, but they can only be purchased from a licensed retailer or directly from the manufacturer. If conditions are very bad the coccidiostat in the food will not prevent the disease, and outbreaks will need to be treated with antibiotics; birds will then generally recover rapidly.

Fowl Pest

Fowl pest, more properly known as Newcastle disease, is unlikely to trouble the average bantam keeper, but as it is one of two 'notifiable' diseases, it should be mentioned because should it ever re-emerge in the UK it will have an effect on all poultry fanciers. The last major outbreak in this country was in 1997 when it affected only eleven farms, but it swept through Denmark as recently as 2003, affecting 135 flocks, and it had immense economic repercussions within the country's poultry industry. It is another respiratory infection, the symptoms being typically a long

A small coop is useful in housing a bird that looks a little 'off colour'.

gasping inhalation, mucus from the nostrils, and greeny-yellow diarrhoea. It is, however, possible to vaccinate, and I well remember injecting my own stock during a national outbreak in the early 1970s.

Fowl Plague (Avian Influenza)

This is the second of the two notifiable diseases, and DEFRA is concerned that owners of small flocks are made aware of its potentially devastating effects following the virulent strain that affected the Netherlands, Belgium and Germany in 2004.

Commonly known to be carried by migrating wild birds, it is theoretically possible for both fowl plague and fowl pest to enter a smallholding. Possible symptoms are: birds standing around in isolation; wings and tails drooping; drinking a lot but refusing food; fast, laboured breathing; and a very high temperature, which drops sharply to subnormal immediately before death. There is no treatment or vaccine.

Infectious Bronchitis

This disease is more likely to be seen in smaller units, as commercial flocks are generally vaccinated as a matter of course. The virus, caused by a coronavirus, is airborne and transmits from bird to bird via nocturnal coughing and sneezing. Except for very young birds, the mortality is quite low, but poultry could succumb to secondary infections. It is possible to protect via vaccination.

Marek's Disease

Sometimes known as fowl paralysis, Marek's can occur as a result of stress or even an abundance of parasitic worms. It used to be a major contagious disease in domestic poultry, but is nowadays less of a problem due to vaccination; it is thought that there could well be a genetic predisposition. Caused by the herpes virus, lesions can form in birds as young as six to eight weeks of age. In its 'classical' form, signs show as lameness in one or both legs, which develops into a general paralysis. Affected birds often lie with one leg forwards and the other stretched backwards, though they seem otherwise alert and happy to eat food if this is put in reach beside them. Birds may later suffer from a bunching of the claws, a gradually worsening wing droop, and possibly even a twisted head or paralysed neck.

Although repeating oneself risk's being irritating, it is nevertheless well worth re-iterating the need to be constantly on the lookout for signs that your birds are not completely healthy and happy. Hygiene goes a long way towards ensuring that they are, but at the first indication of disease, it is important to get an accurate diagnosis and act promptly.

OPPOSITE: Black Japanese male. Photo: Rupert Stephenson

CHAPTER 8

Bantam Ducks

Despite being a lover of all types of fowl – I even once kept a flock of guinea fowl, which had the very noisy habit of running up and down a corrugated tin roof close to my bedroom window first thing in the morning – I have to admit to being less than fond of domestic ducks. To my mind they are intolerably messy, turning a pristine piece of grass into a messy quagmire within hours – but even I cannot help being charmed by the attractiveness and character of bantam ducks.

Large species of duck can be messy unless you have space and water available.

Can you have such a thing as bantam ducks? Well, as we have seen in the earlier chapters there is a world of difference between 'true' bantams and those breeds of domestic poultry that have been 'bantamized'. Rather like bonsai trees, which are artificially dwarfed versions of the natural stock, there is no reason why any breed cannot in theory be miniaturized, and advertisements for bantam ducks are often seen in the specialist poultry and smallholder magazines. The Domestic Waterfowl Club of Great Britain (originally formed to promote and encourage people to keep all breeds of domestic duck) classifies the Black East Indian, Miniature Crested, Miniature Silver Appleyard and Silver as being true bantam ducks, but perhaps the original true bantam is the Decoy or Call duck. Enthusiastic breeders have bantamized other breeds, and the general fancier may be tempted to try his or her hand at keeping a few as pets.

CHOOSING A BREED

It can be easy to confuse ornamental breeds of duck such as Carolinas and Mandarins with 'bantam' types, purely on account of their small size, but technically they are classified as wildfowl rather than domestic (although the American Bantam Association does actually recognize them in their bantam duck classes). There isn't quite the same scope for choosing a particular type of bantam duck as there is when it comes to selecting a poultry breed; nevertheless, some versions of our domestic duck – all of which, with the exception of the Moscovy, originate from the wild Mallard – are very attractive, and it may be possible to find bantamized examples of the following: Aylesbury, Bali, Blue Swedish, Buff, Campbell, Cayuga, Crested, Hook-billed, Indian Runner, Magpie, Orpington, Pekin, Pomeranian, Rouen, Saxony, Swedish, Welsh Harlequin.

I have deliberately omitted the Call duck from the list, as it is well worth a special mention in a chapter of this nature. Originally referred to in earlier writings as the Decoy, its name changed to the Call at some time in the latter half of the 1800s, when the only known colours were mallard and white. Despite originating in the East Indies, the Dutch long ago realized the value of the Call duck in encouraging native wildfowl to their breeding sanctuaries. Their ability to call and decoy large numbers of breeding stock of Mallard into designated areas resulted in as many as 2,000 young ducks being bred on just one 6ha (15 acre) lake (*Game Conservancy Annual Review*, 1953).

Today they are the smallest and most popular of our hobby breeds, as they are very tame and easy to rear. In relation to their size they seem particularly noisy, but that only adds to their character, as long as you are situated far enough away from any neighbours who may complain. What they lack in laying capacity the females make up for in being superb mothers, and they will brood and rear their ducklings to maturity with very little assistance – perhaps another reason why they are an ideal choice for the first-time fancier. There are several varieties of plumage, the most common being white, but all colours are seen, ranging from apricot to yellow-bellied.

HOUSING

There are some who say that ducks require no housing, especially if they have access to a large pond, where they should be reasonably safe from predators at night. To my mind, however, it is essential to give them at least a 'home base' in which they can shelter, feel secure, and lay their eggs. It need not be too elaborate: its construction should be within the capability of even the most ham-fisted of DIY exponents! There is certainly no need to provide them with all the considerations of draught proofing and lighting that is so necessary for the well-being of poultry.

A south-facing aspect will prove advantageous, but basically all that is required is a

Call ducks: all-'mallard' type, with the exception of the 'apricot' to the side of the photograph.

Provided that it is draught-proof and well ventilated, most types of shed can be adapted for use by bantam ducks. It need not be too elaborate and can, in fact, be quite attractive.

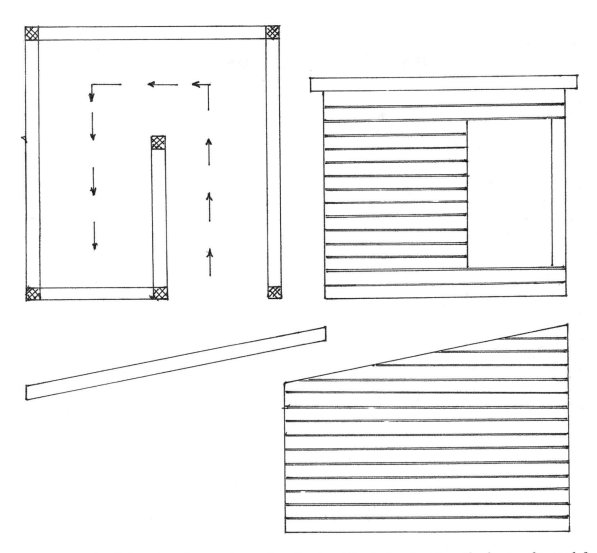

Plan, front and end elevation of a magpie-proof nesting box. The duck enters from the front and turns left into a more secure nesting chamber. The roof is easily removable for access. Dimensions are roughly 45cm (18in) ¥ 30cm (12in).

dry, well-ventilated house with a raised floor. Ideally, the floor should be constructed of wooden slats, rather than of solid boarding, as this will help the chosen litter to drain, and prevent the interior from becoming damp and smelly. It is a wise precaution to fix the underside of the flooring with wire netting in an effort to deter rats. Because the floor is slatted, it will obviously be impossible to use some of the previously recommended forms of litter such as wood shavings, and you will have no alternative but to resort to good quality straw or dried bracken. Loosened daily and changed weekly, however, fungal spores will not have the opportunity to develop in the litter, and the health of your ducks should not be at risk.

If you want to include nest boxes, keep them simple (you may well find that the ducks will just lay in the straw litter anyway). For bantam ducks kept on a pond and within the safe confines of a fox-proof fence similar to that described in Chapter 3, small nesting boxes situated around the water's edge will provide ideal sites for egg laying: 22ltr (5gal) plastic drums lying on their side with one end cut open and one third buried prove very successful. This style may,

however, encourage predation from egg thieves such as magpies and crows, who soon learn that an easy meal is available. A more elaborate 'magpie-proof' nest box can easily be made up out of a few scraps of wood, and consists of a box that the nesting duck enters from a front opening, before turning left into a more secure, secondary area.

As ducks do not roost, there is no need to provide a house of any great height in order to accommodate perches. A 'rabbit hutch'-

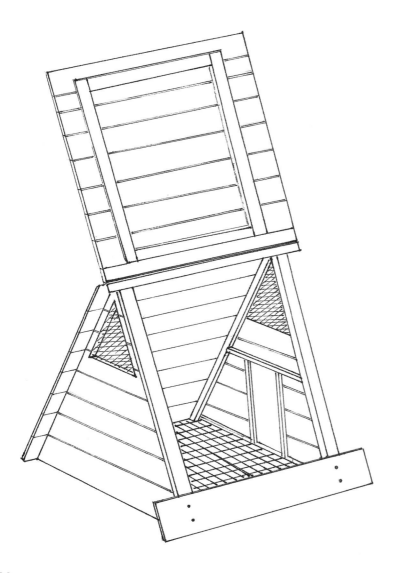

An easily constructed 'A'-shaped shelter suitable for bantam ducks. The floor should be fitted with wooden slats or weld-mesh to keep the interior more hygienic. A covering of straw makes a comfortable bed. Ensure that there is adequate ventilation in the eaves.

shaped house need only be around 1m (3ft 3in) at the front, dropping to roughly 75cm (2ft 6in) at the back; but it is even simpler to construct an 'A'-shaped shed. With either design, a little extra height at the eaves will give better air circulation, especially if the ends of the house are not boarded quite to ceiling level, and instead are merely protected by the addition of strong wire netting or weld-mesh.

The 'rabbit hutch' design should have a door at the front, whilst the 'A' shape more

Bantam ducks may need shutting in at night. A useful addition to an ark and run would be the inclusion of a length of nylon cord, which allows the pop-hole to be opened and closed without having to enter the pen.

commonly has a door at one end. Either way it may pay to fix the door so that it is completely removable, which will help in the weekly clean-out and in keeping the interior smelling more sweetly in warm weather. It has been suggested that a drop-fronted door can double up as an access ramp. (Bantam ducks, being less agile than their poultry cousins, will definitely need some form of ramp, especially if the shed is raised any distance from the floor.) In practice I have found that drop-fronted doors, or indeed sliding ones, are rarely a good idea, as the runners or hinge joints soon become full of debris and refuse to operate efficiently. For ducks, I would suggest that the best solution is a door that is completely removable, held in place at night by four 'button' catches.

Whether or not your ducks need a wire-netting run depends entirely on the garden and security from the neighbour's pets. Unlike bantam fowl, bantam ducks have no tendencies towards scratching about in the flowerbeds and can actually prove quite useful in eradicating unwanted pests from the garden. They are especially fond of slugs and other 'beasties' that seem intent on doing damage to the vegetable patch, which has to be a definite plus point. I began this chapter by dismissing large ducks as being messy creatures, but the small bantam varieties are certainly less guilty of this, and will actually add colour, movement and interest to an enclosed, and therefore safe, garden.

Before giving your newly acquired pets total 'free range', I would suggest penning

Limited space, breeding programmes or the danger of worrying by neighbouring dogs may make it necessary to keep ducks in a movable house and run.

them for at least a fortnight. This will teach them where home and food are situated, and will also help in training them to return to the house at night. They will probably need encouragement every time: like naughty children at bath time, ducks are reluctant to go to bed. Any 'shepherding' or 'herding' necessary must be done with the minimum of fuss as ducks tend to be more excitable than other forms of fowl, responding best to gentle coaxing and kind words.

Unlike other breeds of bantam, the Call duck is an excellent flyer and may well need wing clipping in order to prevent its escape. When buying new stock, remember to enquire as to whether they have been clipped, and if they have not, ask the seller to show you how to do it. Pinioning, which involved the permanent removal of the manus, or outer part of the wing, and was usually done in the very early stages of a duckling's life, has been the subject of much heated debate over the last couple of years. At the time of writing, it looks likely to be made illegal in domestic circumstances. In America, a tendon-cutting technique is sometimes advised, but a percentage of birds will develop wing-droop, which is obviously to be avoided in bantam ducks kept either for pleasure or exhibition. Clipping the primary feathers of one wing will keep ducks grounded until their next moult, although you may well find that, as time goes on and your pets settle in, it becomes an unnecessary practice.

With limited space or marauding household pets, there is no reason why any bantam ducks should not be housed with the added security of a run attachment, and provided it

A small, purpose-built pond, ideal for bantam ducks. Some method of drainage should always be included so as to facilitate easy cleaning.

is moved on to fresh ground regularly, neither the garden nor the ducks will suffer. In particularly cold areas, it might be a good idea to at least partially cover the pen sides with plastic during the winter months and thereby cut down the effects of wind.

FEEDING MATTERS

This section begins by discussing water rather than food, since as everyone knows, ducks take to water, well, like ducks to water! Although a pond is not essential, the very least requirement is a 'paddling pool' deep enough for them to submerse their heads. (It is important that ducks keep their eyes and vents clean.) The water source should be easy to drain unless it is supplied by a constant flow of running water, and thought must be given as to whether a small duck can squeeze out of either outflow or inflow – a brick or breeze block is often all that is required.

Drinking water is best given in troughs or fountains, into which the ducks cannot actually have access, otherwise they will just treat it as another pond. It needs to be supplied in both the house and run, because unlike poultry, ducks eat and drink at night. It also needs to be regularly changed, or it will be quickly become fouled. Ducks of all types like to eat and drink alternately, often actually dunking their food in water before eating. The container in the house is best raised on a slatted or wire netting-covered platform to prevent the surrounding litter from becoming too wet. Unlike the platform described earlier as being ideal for bantam fowl, there is no need to protect the underside with a water-catching tray, as most of what is spilt will soak away through the litter and 'open' floor.

Feeding is as simple as housing. Bantam ducks will thrive on a basic diet of crumbly wet layers' mash and a separate feed of grain in the evening. Beware of providing more mash than the birds can eat in one session, as moist mash has the unfortunate habit of quickly becoming stale and sour, especially in

warm weather. There is no reason why small poultry pellets should not be fed, but ask your supplier to show you a sample before committing yourself to the expense of a full bag and then discovering that the type you have chosen is too large. The ideal is a well-balanced waterfowl or game-bird ration, but make sure that the food is additive free: for some reason, anti-coccidiostats, for example, are detrimental to the well-being of ducks. If you are beginning your collection of bantam ducks by hatching a few eggs under a broody chicken, then the youngsters will get all the nutrients they require from ordinary chick crumbs.

Confined ducks will benefit from being given a few extra treats, as they are unable to free range and thus supplement their diet. An enthusiastic and very knowledgeable breeder of bantam ducks whom I contacted in connection with this section advised giving the occasional earthworm, mealworms, slugs and peanuts – but he also warned of a fungus sometimes found in badly stored peanuts, which can affect the health of ducks in particular and poultry in general. Cooked vegetables and brown bread are appreciated. It is also important that a supply of mixed grit is included, as once again, confined birds are unable to source their own.

BREEDING AND HATCHING

Much of what has been written in Chapter 5 concerning the breeding, rearing and hatching of bantam fowl applies equally to bantam ducks. It has already been pointed out that all ducks (with the exception of Call or Decoy birds) make bad mothers, and so it is likely that any hatching will be done by broody bantams, chickens or possibly in small incubators.

Albeit that Call ducks are unique in their mothering instincts, it is well worth noting that if the intention is to let them hatch-off their own brood, the eggs should not be removed from the nest. Laying just one clutch in a normal year, the female needs to be left

Call ducks will brood their own eggs very successfully. A nesting box such as the one shown is ideal.

with all the eggs otherwise she may not sit. Some breeders of Call ducks tell me that it is likely to be late in the season before the duck begins sitting, and they have also observed that fertility often tends to be low. There could be two reasons for this: the fertility of any eggs, whether chicken or game, becomes less as the season progresses, or – and this option is considered to be the most likely – low fertility in Call ducks could be a result of most stock being inter-related.

The majority of bantam duck fanciers will probably breed from either a pair or trio of birds, but some drakes can be a little aggressive towards the females during the mating season. This problem is more likely to occur when there is competition between males, either in the same pen or when housed next door to each other. In the event that it is necessary to change a mating, it can be done most easily if the former mates are not able to see each other. With just one pen, a drake is

It is important to choose the heat source carefully when rearing young ducklings. Water splashed on to lamps will shatter them, and it is better to use dull emitters with a separate light source. Photo: Geoff Burch

usually quite happy with either a single female or a harem of up to five.

Hatching any kind of duck egg requires a little more moisture than when incubating poultry, so it is important to remember to add some aired water in and around the nest, and also to spray or splash some on to the broody spot of the would-be foster mother. Eggs set in incubators need special attention, and in dry regions, spraying daily with warm water from the tenth day of incubation has proved beneficial: where atmospheric humidity is normally high – in areas of above-average rainfall – any additional moisture is probably not necessary.

When using a small incubator for hatching, it is never a good idea to mix duck eggs with any other types, not just because of the differing hatching times, but also because a mixture of eggs will affect the temperature and humidity of the incubator at hatching time. Dirty eggs should be cleaned with wire wool before being incubated, and very soiled ones may need to be washed using a proprietary egg sanitant to reduce the chances of infection.

It is obviously far easier to hatch one's bantam ducklings under a broody hen, but if there is no alternative, artificial brooding works well. Remember they do not need as

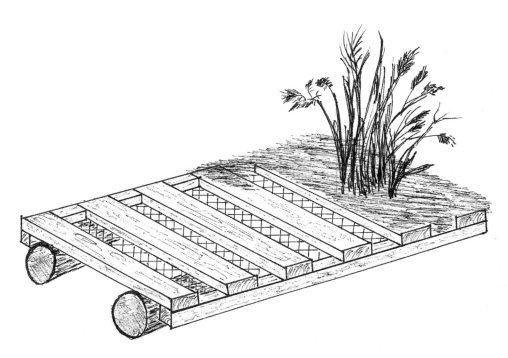

Once mature, bantam ducks are quite hardy and can be left to their own devices. Where a small pond is available, it is sometimes possible to construct an artificial island on which they can spend some of their time. Made from telegraph poles or drums (as floats) with battening attached both across and lengthways, the base is easy to assemble. Wire netting and some wads of straw are all that is required to complete a platform, and once sodden, the straw will provide an ideal medium in which to plant vegetation. (As the straw breaks down quickly, there is little or no risk of fungal diseases.) Such a construction will, of course, need to be anchored by weights or ropes attached to pegs on the bank-side.

high a temperature as domestic fowl, and a heat pattern of around 30°C (86°F) is ideal for the first few days. Avoid using an infra-red heater for ducklings – being messy in their drinking habits, there is a good chance that they will cause the lamps to blow by splashing water on them. It is far better to use the dull emitters, as they are more robust and will not shatter.

For birds supposedly at home in an aquatic environment, it is surprising to see that water causes potential problems in some aspects of duck keeping, not least when it comes to rearing. The chosen drinker for bantam ducklings must not be too deep, as they are notorious for 'playing' in water and by doing so, suffer from chilling. If a narrow-based fountain is unobtainable, pebbles, a length of hose, or even marbles placed around the base will prevent full access, and should go a long way towards preventing any deaths. For the same reasons, it is important to protect them from rain until their mature feathers develop at about three months. (At the other end of the meteorological scale, young ducklings can suffer quite badly from sunstroke, and great care must be taken when they are first placed outside. Make sure they have access to shade at all times.)

Artificially reared bantam ducklings will appreciate the addition of fresh greens to their diet until the time they are old enough to have access to a run in the garden. Small quantities of freshly cut grass or chopped dandelions are ideal, but avoid giving them too much at any one time otherwise it will quickly become part of the litter, rather than being eaten.

SHOWING AND EXHIBITION

There are the same opportunities to show bantam ducks as there are to exhibit bantam fowl, the major difference being that there are fewer ducks being shown. This is a shame, and even though most people keep them as pets for their character and charm, it is well worth considering the showing aspect.

In an average year, around five or seven major shows are held, and these may attract between eight hundred and a thousand exhibits; but on a smaller scale, there are usually 'waterfowl' sections included at all shows, so there should be no difficulty in finding a venue if showing appeals to you. If you have never experienced a poultry show of any kind before, it is obviously a good idea to go along as a spectator before filling in your first entry form.

As with every type of showing, no matter whether it is cattle or flowers, there are certain standards to which your chosen breed of duck must conform. Bantam duck breed standards can be found in the book *British Poultry Standards*, the poultryman's 'bible'.

Judges for the bantam duck sections are often selected from the judges' panel that features in the *Poultry Club's Year Book*. The way that judges are selected is explained in greater detail in Chapter 6, but it is necessary to repeat some aspects briefly in order

Careful examination of the 'top' birds will prove educational, as will close scrutiny of the various breed standards. Photo: Francesca Hobson

Mallard Call Duck

Bantam Appleyard

White Indian Runner

Welsh Harlequin

Some breeds of bantam duck will be immediately recognizable by their shape, build and stance.

to make sense of how individuals become qualified to judge this particular section. The list that appears in the *Year Book* is divided into four categories: panels A, B, C and D. To gain promotion up the ladder to the level of a grade 'A' judge, a total of seven examinations have to be taken. One of the seven is the 'Waterfowl Test', which is organized in conjunction with the British Waterfowl Association. Candidates have the option of specializing by taking four separate tests – geese, heavy ducks, light ducks and bantam ducks – which when taken individually, will qualify a judge for Panel C level and also BWA and Poultry Club shows.

It is important for any bird to be in good health and prime condition if it is to have any chance of winning – an underweight specimen will not possess the smooth form and overall appearance that will catch the judge's eye. Bantam ducks are prone to poor skin condition, and during the winter in particular, dead skin can accumulate on feet, legs and bill. These areas should be attended to well before the beginning of the show season, and experienced exhibitors often advocate the application of Vaseline or other natural moisturizers.

Damaged feathers will obviously ruin a duck's chances in the show pen, and any broken or soiled feathers must be removed in plenty of time for them to re-grow. Provided that birds are kept in clean quarters for at least a week before the show date, and that they are provided with clean bathing water daily, they should preen and oil their feathers into superb condition. A successful poultry-showing friend of my grandfather's used to swear by the use of a silk handkerchief to wipe down the feathers, and it certainly provided his particular exhibits with an extra sheen to their plumage.

If your attempts at showing are unsuccessful and you are convinced your ducks conform to the standards required, do not be afraid to seek out the judges of the day and ask them to explain their reasoning. It is the only way to learn.

Sometimes it is possible to 'show off' one's stock without putting them under the close scrutiny of a judge. The organizers of many of the agricultural shows and country fairs held each year like to include and promote any livestock clubs in their particular region. As a member of the local poultry fanciers' group, there is often the chance to participate in creating such an exhibition, which is not only good public relations but also a perfect place to meet fellow enthusiasts. Poultry keeping of any kind can sometimes be a solitary hobby, and the opportunity to swap ideas and advice is one not to be missed.

HYGIENE AND HEALTH

Bantam ducks are easy to maintain, and provided they are cared for sensibly, suffer from very few of the health problems that can occasionally affect commercial flocks of domestic duck. It is unlikely, for example, that they will ever succumb to duck virus enteritis (despite the sudden outbreak in 2003 that affected several breeders throughout the country), or hepatitis, or anatipestifer infection, and so to outline in detail all the symptoms would be wrong in a chapter of this nature – it would be like reading a 'home doctor' book and becoming immediately convinced that one has every disease known to man!

Although it is necessary to be aware of potential problems, there is a very real danger of thinking the worst. One of the symptoms of anatipestifer infection (duck septicaemia), for example, is watery eyes and nostrils – but in all probability 'weepy' eyes are more likely to be a result of a hot, dry summer. Because of the excitability of ducks, there is a slight chance that they may cause temporary damage to their thighs or legs if rushed during their daily routine, or are suddenly frightened by the appearance of a household pet or fox. Again, it is important not to assume the worst and think that lameness is a symptom of something more serious.

A 'Sussex' ark makes a good home for bantam ducks, but the ramp should be fixed at a very gentle slope to avoid possible leg and thigh damage.

Bantam ducks are nothing like as prone to ectoparasites as poultry, but nevertheless, keep an eye out for worms, mites, lice and fleas. Whilst not life-threatening, they will certainly affect the bird's appearance, behaviour and general well-being. Treat them in the same way as for any other types of poultry.

Parasitic fungi can also attack the respiratory system, and with mouldy straw and damp grain being a frequent source of such infections, cleanliness is of paramount importance. The breeders of bantam ducks have told me many times that outbreaks of disease generally only occur as a result of bad management, and that if good food is provided, and overcrowding, bad ventilation and dirty conditions are avoided, it should be possible to enjoy a lifetime of bantam duck

keeping without ever encountering disease in any shape or form.

A regular worming programme is, however, a wise precaution. Most breeders add a wormer to their food twice a year, once in the spring just before the breeding season, and once in the autumn. Despite the warnings earlier in this chapter to ensure that proprietary feeds are 'additive free', it seems that the inclusion of a wormer such as 'Flubenvet' will cause them no ill effects.

IN CONCLUSION

Like bantam fowl, it seems that bantam ducks are the perfect answer for the enthusiast without a great deal of space at his or her disposal, or for those like me who happen to think the smaller varieties of poultry have

Bantams or bantam ducks? If space permits, why not have both?

more charm and charisma than their larger cousins. They are the solution for anyone with a hankering for a waterfowl collection, and tend to be more ornamental as well as less messy than some of the larger breeds of duck. Easily cared for, virtually disease free and kept correctly, bantam ducks require little or no special preparation for the show pen – even a single pen could prove to be just the beginning of a lifetime's fascination and pleasure.

But then, that's bantams for you, whether they be duck or fowl; as Mrs Ferguson said, all those years ago: 'Gems of beauty ... impudent ... (and) captivating.'

A Dictionary of Bantam Terms

To include all the technical poultry terms would take up most of this book. It is my intention to concentrate only on explaining definitions likely to be encountered in further reading, or when talking to experienced bantam fanciers and exhibitors.

addled   A fertile egg, the embryo of which has died during incubation.

airsac   Air space found at the broad end of the egg. It can be seen when candling and its size denotes the freshness and, during incubation, the development of the embryo.

AOC   Showing term, normally seen on a schedule. An abbreviation for 'Any Other Colour'.

AOV   As above, but meaning 'Any Other Variety'.

ark   Small, portable bantam house.

axial feather   Short wing feather found between the primary and secondary flight feathers.

barring   Plumage markings of equal space and width across a feather. Seen distinctly in Barred Wyandotte, but less well defined in the Maran breed.

beard   Tuft of throat feathers; also called *Muff*. Seen in breeds such as Faverolles, Houdans and some varieties of Polish bantams.

bloodspot   Seen in the laid egg; the reasons for its appearance could be due to generic, nutritional causes or the hen experiencing a sudden fright.

boots   Feathers growing down the legs and across the toes. Seen most obviously in Brahma and Cochin breeds.

boule   Type of feather formation found on the necks of some continental breeds.

brassy   When the feathers of the neck and back of light- or white-coloured birds develop straw-coloured areas. This is caused by sunshine or extreme weather. A great problem to the showing enthusiast, who normally keeps white, lavender and other light-coloured birds in the shade.

breed   Any group of fowl whose characteristics distinguish it from any other and which are transferred to subsequent generations.

breeding pair/trio   Should be a male and the appropriate number of females.

brooder   An artificial heater for rearing young birds.

broody As an adjective, having the natural instinct to sit on eggs once a clutch has been laid. As a noun, a hen bird showing the same inclination, which will sit either on her own eggs or those substituted.

caecum One of two intestinal pouches found at the junction of the small and large intestines.

candling Using a source of light behind the egg to detect freshness, hatchability and flaws in the shell structure before incubating. During incubation, eggs can be candled to check fertility and humidity.

cape Feathers running from the back of the head down to the shoulders. The capes of cock birds are much sought after by fly fishermen, and surplus cocks could be killed at maturity for the table and their capes sold.

carriage A show term denoting the ideal stance of a particular breed.

chalazae 'Shock absorbers' of albumen around the yolk of an egg. It is important to let these reconstitute themselves after eggs have been travelling for a long distance before setting them under a broody or in an incubator.

china eggs Sometimes referred to as 'pot eggs'. They are used to encourage young birds to lay in the nest boxes, and are also placed under a broody hen to settle her in to the sitting coop before replacing them with fertile eggs

clears Infertile eggs found after candling.

cloacae 'Collection' point for the bird's excrement before it is finally evacuated.

cobby Term used to denote a short-backed, rounded body. An ideal specimen would fit in a circle drawn around the body. A show requirement of some breeds of which Wyandottes are probably the prime example.

cock A mature male after its first breeding year.

cockerel Male bird before it is known as a cock.

cock breeder *See* Chapter 5.

comb Generally the horny muscle that appears on the heads of most breeds. Specifically, there are several different types:

> *cushion*: almost circular (Silkies)
> *horn*: two spikes appear at the top of the comb (La Fleche)
> *leaf*: (Houdan)
> *mulberry*: same as '*cushion*'
> *pea*: three small combs lying parallel, with the centre one forming the highest point (Brahma)
> *raspberry*: like half a raspberry! (Orloff)
> *rose*: broad, solid comb, nearly flat on top, covered with several small regular points and topped off with a leader or spike (Rosecomb, Sebright)
> *shell*: same as '*leaf*'
> *single*: two types, '*large*' (Ancona) and '*folded*' (Leghorn). Probably the most common type of comb seen amongst the various breeds of bantams.
> *strawberry*: like half a strawberry! (Malay)
> *triple*: same as '*pea*'
> *walnut*: same as '*strawberry*'

coop Small hutch usually used to house a broody hen when sitting and/or her chicks.

coverts Wing coverts are feathers covering the top of the flight feathers; tail feathers are at the root of the tail.

crest Tuft of feathers on the heads of some breeds. Sometimes known as a 'top-knot' or 'tassel'.

crop Food collection sac at the internal base of the neck. Here food is 'softened' before being passed into the digestive system.

crossbred 'Mongrels' or hybrid bantams. Will not breed pure to any type.

cushion Feathers over the hen's back near the tail. Can be a show fault in some breeds.

debeaking Trimming back the bird's upper mandible to prevent feather or vent pecking. It should not be necessary with contented bantams, and if practised, would obviously prevent birds from being shown.

double lacing Term to describe plumage pattern. Best example is probably the Indian Game.

double mating *See* Chapter 5.

dropping board Removable board fixed under the perches to catch excrement.

dubbing It used to be common practice to cut back the comb and wattles close to the head, especially in Old English Game varieties. Had its origins in cock fighting.

duck-footed A serious show fault where the rear toes are out of line.

duckwing Colour description of several breeds.

ear lobe May be red, white, blue or purple depending upon the breed.

'faking it' A deliberate attempt by an exhibitor to deceive the judges by dying feathers or covering up faults on his bird.

feathered legs Descriptive term for breeds that have feathers covering the legs to a greater or lesser degree. *See also* 'boot' above.

flight feathers The large primary feathers on the last half of the wing.

fold unit Portable combined house and run.

force moult Artificially persuading a bird to moult at an unnatural point of its cycle. Sometimes carried out to ensure a bird is in prime condition for a show on a given date; however, it is not, in my opinion, to be recommended.

fountain Drinking vessel with a reserve supply of water. Traditionally ceramic or galvanized, but increasingly made of plastic.

foxy Show fault denoting unwanted red colour in the plumage of some breeds (usually in the wing area).

Frizzle A breed that has feathers twisted and turned back in the opposite direction. '*Frizzled*' can mean a fault in other breeds and is usually associated with poor plumage.

furnished A fully feathered bird. The expression is most commonly used to describe a cockerel with full tail and sickle feathers.

gizzard Grinding stomach with muscular lining for pulping food.

grit *See* Chapter 4.

ground colour Like the base of an artist's painting, the term is used to describe the main plumage colour on which other markings are noticed.

gullet Sometimes known as the 'oesophagus', it is the tubular structure leading from the beak to the crop.

hackles Long, pointed neck feathers. Also found on the saddle, where the feathers are rounded in the hen and pointed in the cock. (The latter are sometimes known as 'hangers' on a cock bird.)

hard-feathered Terminology to describe category of feather type – often seen in show schedules. For a full explanation *see* Chapters 1 and 6.

headgear Comb, wattles and ear lobes.

heavy breed Describes those breeds whose ancestry possibly derives from the Cochins and Brahmas. Despite the name, the Light Sussex is a heavy breed!

hen Female after her first laying season.

horn Beak colour shadings. Especially noticeable in the Rhode Island Red.

in-breeding When members of the same family are bred from, through several generations. Experienced breeders occasionally practise it in order to re-establish a good point, but it is far better to try and rectify the fault by using a good bird from unrelated stock.

keel Bony ridge of the breastbone.

lacing Strip of a different colour around the edge of the feathers.

light breed Usually Mediterranean in origin, where their light bones and quick feathering makes them adaptable to hot weather. Generally good layers, but somewhat flighty if startled.

line-breeding An understanding of basic genetics is useful before undertaking line-breeding, literally within the family line but without '*in-breeding*', which is to be avoided at all costs.

marbled Sometimes known as *mottling*, both indicate spotting on the plumage of bantams such as Anconas.

marking Any lacing, barring, spangling or pencilling on the plumage.

moons Spangled markings on the plumage.

mossy A defect in the colouring of pure-breds, indicating indistinct markings.

moult The period when a bantam sheds its old feathers and re-grows new ones. Generally occurs in the late summer/early autumn.

muff (*or muffling*) The beard and whiskers found on some breeds, such as Barbu d'Anvers and Faverolles.

oil sac When a bird is observed preening itself and continually wiping its beak up the base of the rump, it is sourcing the natural oils found there in the oil sac or gland. Essential for feather condition.

out-breeding Opposite of '*in-breeding*'! Basically, mating different lines of the same breed. *See 'line-breeding' above.*

pencilling Small marks or stripes on the feathers. Can be straight across (Hamburg), slightly V-shaped (Campine), or crescent-shaped following the outline of the feather (Brown Leghorn). Sometimes also known as 'bands'.

peppering Expression used in the show world to indicate irregular spot markings. A fault in some breeds.

pile Colour description of several breeds.

pin feathers   In the shooting world, highly prized feather taken from the wing of a woodcock and often used as a paintbrush by artists; however, the term is used generally amongst poultry fanciers to describe the new feathers emerging after the moult.

pinioning   Sometimes done to bantam ducks to prevent flight. *See* Chapter 8.

primary feathers   The long, still feathers at the outer tip of the wing (there should be ten of them).

pullet   Young hen from hatching to the end of the first season.

pullet breeder   As with 'cock breeder', *see* Chapter 5.

recessive   Genes which may not be evident in the initial mating, but which may well manifest themselves in subsequent generations.

roach back   A deformity of the vertebrae showing as a hunched back.

saddle   Posterior part of the back of a male bird, equivalent of hen's 'cushion'.

scales   Found on legs and toes.

secondary feathers   Quill feathers on the wing, which are usually visible when the wings are folded or extended.

self-colour   Plumage of the same colour throughout.

shaft   The stem or base of any feather. On newly moulting birds, blood can often be seen internally through the shaft.

sickles   Long curved feathers on the outer sides of a cock bird's tail.

sitting   As in 'I bought a sitting of eggs': usually nine or thirteen in number, depending upon the size of the broody hen being used. Could also be termed a 'clutch'.

soft-feathered   Terminology used to describe category of feather type – often seen used in show schedules. For full explanation *see* Chapters 1 and 6.

spangling   Plumage in which there is a different colour towards the tip of the feathers (Hamburgs, Sebrights).

split wing   A deformity of the wing that shows itself when the axial feather is missing. It is thought to be hereditary and it is therefore best not to breed from any bird displaying this fault.

spur   The pointed, horny projection at the base and rear of a cock bird's legs. Small nodules are sometimes noticed on the female. The spurs on a cock grow longer as it matures, but it is not, as is sometimes thought, a reliable indicator of the bird's age.

squirrel tail   A fault sometimes seen in heavy breeds where the tail feathers grow towards the neck. Bantams with this defect should not be bred from.

strain   A group or flock of bantams carefully bred over several generations by an individual fancier. Quite often, the strain will be known by the name of the breeder. Some producers of particular breeds may well have separate 'cock-breeding' and 'pullet-breeding' strains. *See* Chapter 5.

tail feathers   The stiff, straight feathers of the cock bird normally found under the sickle feathers. On some breeds, however, notably the Cochins, Brahmas and Orpingtons, they are not present.

trachea   Otherwise known as the windpipe and is the part of the respiratory system that allows air to pass from the larynx to the lungs and bronchi. Sometimes the trachea can be affected by dust and worm infections. *See* Chapter 7.

type   Bantams with the correct confirmation for their particular breed.

undercolour   Colour of the downy part of the plumage that is normally only seen by gently brushing back the outer feathers.

variety   Bantams of the same breed but which are different in colour. Everything else, for example the body shape and comb type, remains the same.

vent   Rear 'opening' through which droppings and eggs are excreted.

vulture hocks   Seen on some feather-legged breeds, the term 'vulture hocks' describes stiff feathers showing on the elbow or hock area of the leg.

wattles   The folds of skin hanging on either side of the lower beak. Ideally, they should always be of equal length, and if not, would be marked down by most judges.

weathering   *See* 'brassy' above.

wingbow   Upper or shoulder part of wing.

wing clipping   Sometimes it is necessary to clip the primary and secondary feathers of one wing to prevent the lighter breeds of bantam from flying. The feathers will re-grow in the next moult, but until then it would not be possible to show the birds.

wing coverts   Feathers covering the roots of the secondary quills.

wry tail   Any tail that is carried either to the left or right of the imaginary continuation of the backbone is said to be 'wry-tailed'. It can affect both hens and cocks, and neither should be bred from, as it is a genetic defect.

Index